# POLITICS AND PEOPLE
# IN ETHOLOGY

# POLITICS AND PEOPLE IN ETHOLOGY

## Personal Reflections on the Study of Animal Behavior

Peter H. Klopfer

Lewisburg
Bucknell University Press
London: Associated University Presses

Associated University Presses
440 Forsgate Drive
Cranbury, NJ 08512

Associated University Presses
16 Barter Street
London WC1A 2AH, England

Associated University Presses
P.O. Box 338, Port Credit
Mississauga, Ontario
Canada L5G 4L8

The paper used in this publication meets the requirements
of the American National Standard for Permanence of Paper
for Printed Library Materials Z39.48-1984.

**Library of Congress Cataloging-in-Publication Data**

Klopfer, Peter H.
    Politics and people in ethology : personal reflections on the study of animal behavior / Peter H. Klopfer.
        p.    cm.
    Includes bibliographical references and index.
    ISBN 0-8387-5405-8 (alk. paper)
    1. Animal behavior—Research—Political aspects. 2. Ethologists. 3. Klopfer, Peter H. I. Title.
    LQL751.K594      1999
    591.5'072—dc21                                                    98-28140
                                                                              CIP

PRINTED IN THE UNITED STATES OF AMERICA

*To Martha*

# Contents

CONTENTS

# Foreword

In the late fifties, I visited the Madingley Field Station for Animal Behaviour of Cambridge University. There Robert Hinde introduced me to the several graduate students then working with him and Bill Thorpe. One of these was Peter Klopfer, a fellow American. Since that time, I have not seen him once and have had word of him only occasionally from mutual friends at Duke. As I read this book, I came to be all but convinced that he and I have been in continual friendly intellectual and social contact through the years since.

That meeting occurred during the second of my two several-months-long working visits to Niko Tinbergen's ethological lab at Oxford. The first stay, in 1953, was instigated by the seemingly sudden emergence of ethology after World War II, apparent in the United States concurrently with the availability of Tinbergen's book, *The Study of Instinct*. I had just completed, with others, an extensive study of then-current "learning theory," and my own more detailed analysis of B. F. Skinner's systematic approach to the scientific study of behavior.

Both the methodology and the basic concepts of ethology, as I first encountered them, came as a challenging revelation. I had to go, see, and do all I could to learn more about this behaviorism that studied 'organisms' living in their natural environments. So, with the support of the American Philosophical Society, in 1953 I spent a sabbatical working with Niko and his students both in the lab and out in the field, on sand dunes and windswept islands. I also visited at some length the laboratories and field stations of other ethologists, and the laboratories of European animal behaviorists not so readily identified as ethologists, as well as of antiethologists— or, more properly, anti-Nazis (specifically Communists), to wit, J. B. S. Haldane and his spouse, Helen Spurway, who had no doubt that Konrad Lorenz was a thorough Nazi, and ethology a thoroughly Nazi antiscience.

On most of these visits, I was a house guest as well as a visitor to labs.

9

I came to know personally many of the ethologists the reader will meet in this book and to know a good deal about the conceptual framework that guided their research, and perhaps even more about the political context (both personal and ideological) within which they worked.

In the early sixties, my own contacts with the community of ethologists were brought to an abrupt end by "administrative responsibilities." But all my subsequent research (on human behavior) has been profoundly shaped by the methodology and basic conceptual system of ethologists. The principles of ethology were stated succinctly by Tinbergen some years later. Dr. Klopfer has exemplified them through his own career in ethology and restates them for us: The ethologist is concerned with the phylogeny, ontogeny, mechanisms, and consequences of behaviors and investigates all four in their relationship to one another.

The decade 1953 to 1963 was a period when the ideological/political issues confronting ethologists were of far greater—or should I say more heated?—importance than in later years. It was also the period when the "flush-toilet" theoretical views of Lorenz and his statements concerning more striking findings, such as imprinting, became less attractive, as both theories and new discoveries are likely to do. Such was the decade when Klopfer moved from ecology into ethology.

Those ethologists were enjoying their work, having fun, no matter the strong political undertones that were, rightly or wrongly, evident on all sides. New possibilities were everywhere; new findings waited to be made. Alongside the satisfactions of working in a rapidly evolving scientific field, there was some of the suspense of a detective story. These ethologists had personal histories to contend with and fundamental political orientations to question. The Cold War was ignored; the bitter aftermath of the previous war persisted. During meetings, an evening's beery conversation in pub or hotel room was as likely to deal with questions such as whether so-and-so was an SS officer as it was to discuss imprinting or nest relief in the black-backed gull. Questions persisted about what had happened to colleagues on either side of the war.

Dr. Klopfer's book gives us much of the flavor of those times, especially in communicating the sheer enjoyment of their work that characterized the ethologists I worked with and visited. He shows that ethology provided and still provides, not only a scientific career, but what I do not hesitate to term a happy, joyous one.

He also shows us the changes that have occurred over recent decades. As its title promises, his book is a personal memoir. Dr. Klopfer lets the reader meet him and respond to him both as a working scientist and as a person. As a working scientist he is someone who has enjoyed, and continues to enjoy, his work and the people with whom he has been associated, whether as teacher, colleague, or student. As a person he is a man who values both friends and Friends, and he is one with whom one can work, play, make practical jokes, or talk. His good-humored and hearty approach and the physical characteristics that can be inferred from his memoir remind me of none other than Konrad Lorenz, who, with Tinbergen, shaped ethology.

So, I would greatly enjoy and, even more, profit by the opportunity to have chats, long give-and-take talks, with Peter. I'd like to talk about the historical roots, both in biology and in the political arena, of ethology, to discuss the relationship of "hereditary variables" and "eugenic" theory and practice to its methodology, its theories, and its people. I'd like to exchange stories and speculations about the Konrad Lorenz I came to believe I knew well in 1953, the Lorenz who sheepishly identified a small bird that strutted out of the Buldern woods as "My Nazi chicken," and who asked me how I would like to receive unexpectedly in the mail a parcel containing my grandmother's ashes, straight from the euthanizing hospital, as he had his.

I'd like to worry with Peter about the future of ethology, as contemporary ethologists seem progressively to emphasize research on "mechanisms," the "efficient causes" of behaviors, to the neglect of their phylogeny, ontogeny, and consequences for reproduction. We could discuss—or argue about—what has brought so many ethologists in from the field and into the laboratory, with acceptance of the strictures of that narrow concept of scientific methodology, which has produced an experimental psychology that grinds out rigorously scientific trivialities. We might, too, consider alternatives to the current axiomatic wedding of anatomical structure to behavioral function in behavioral research and theorizing.

And I'd like to swap stories and share impressions of the many people whom we've both met and worked with in the field. Who influenced whom, and how? Why didn't Tinbergen accept any of those lush job offers from Germany? How did Hediger wind up in the Zurich Zoo? Who explained to me Germany's defeat in terms of sexy French girls (lost animal juices)? And so on and so forth.

This book is the very stuff with which historians and sociologists of science should work. Without effort, it catches some historians in the very act of rewriting history, which seems to be an ongoing process. It further provides the kind of nitty-gritty evidence that may enable "social constructionists" to develop a more plausible case than they have done. It tells it how it was, and how it is, for one scientist, working in one area.

In 1957, this behaviorist defined *ethologist* as "a behaviorist who likes his animals." That definition was good then, and Peter Klopfer shows that it is still good.

WILLIAM S. VERPLANCK

# Preface

Chance, good fortune, the will of the gods—whatever forces set one upon a particular path, mine have led to the arenas where the study of animal behavior was becoming formalized. It really did not emerge as a separate discipline with its own societies and journals until the 1950s, just when I was commencing graduate studies. For example, there was no U.S. society dedicated to ethology when I was in school—my only professional association with behavior was with the young British Association for the Study of Animal Behavior. Its sister organization, the Animal Behavior Society, began as an interest group within the Ecological Society of America, but it did not appear until the late 1950s. Ethologists were few in number, and the consequence was that we all were able to become personally acquainted. Our informal conferences and casual meetings greatly outnumbered our congresses and publications.

Another circumstance that contributed to my being able to meet and converse with my intellectual colleagues was the post-Sputnik outpouring of funds in support of science and scientists. Those of us fortunate enough to have completed our studies in the late fifties had relatively little difficulty obtaining fellowships and grants. This translated into considerable funding for travel, both for research and to conferences, promoting more and closer ties with one's professional colleagues. I was able to grow up in small-town science, an environment sadly unavailable to the present generation of ethologists.

Frequently I have been asked to commit to paper my recollections of exchanges in those seemingly distant times. At the very least, my younger colleagues have argued, the descriptions of the characters and temperaments of the older ethologists would provide useful insights into how a young science develops.

Should such accounts be written while its actors are still alive? Oral historians can certainly provide a wealth of detail that would otherwise be

13

obscured under time's patina. On the other hand, personal recollections alone are not a history. Memory is selective and participants are in many cases restricted to one of several possible perspectives when interpreting events.

This is not a proper history. It is, rather, a set of reminiscences intended to illustrate the interplay between intellect and personality. Would Konrad Lorenz's views have been as influential had he not been a large, ebullient, charismatic man? Might not Karl Lashley have won a Nobel Prize rather than Lorenz had Lashley been less modest in his pronouncements?

If I have an ax to grind, it is the belief that the science we call ethology, the study of animal behavior, has been influenced as much in its course by the personalities and politics of its purveyors as by their data and the chronology of their development. What I've attempted to do in the pages that follow is to suggest just how particular personalities and particular kinds of politics interacted. I have not attempted to do this in a systematic fashion, but in a personal manner, drawing on my own experiences and interactions with those who have shaped modern ethology (and, of course, how these personalities and my studies have shaped me).

\* \* \*

I thank Prof. D. Wallschläger, my host at the University of Potsdam, FRD, while I was scratching about in the archives, and to my longtime friend and colleague, Prof. K. Schmidt-Koenig (Tübingen, Duke), who acquainted me with several important sources. Helpful critical comments came from L. Klopfer, G. Wing, C. Koonz, K. Schmidt-Nielsen, and D. Smillie, and I thank them. Particular helpful criticisms of earlier drafts came from Nick Thompson, Ethel Tobach, and Kirk Jensen. The tedious task of continuously retyping from my scrawl fell to the lot of Tamra Mendelson, whom I also thank, as well as Armentha Branche, the ever cheerful presence in the office of the Duke zoology department. Finally, I must acknowledge a considerable debt to my sometime running/riding companion, Rachel Toor, whose editorial skills led to my having a heavier wastebasket and a lighter tome than would otherwise have been the case.

# POLITICS AND PEOPLE
# IN ETHOLOGY

# 1

## Personal Beginnings

Since television has begun featuring programs such as *Nature* or *Nova,* the seductions of the study of animal behavior in particular and biology in general are experienced by every child. In my youth, the biology course, either in high school or college, was one's introduction to the life sciences. For a fortunate few, family, friends or fortuitous circumstances provided an extracurricular introduction. For most others, this first formal encounter with academic biology was the last. In my experience, certainly, nothing was better suited to discourage an interest in biological problems than first-semester biology.

Admittedly, my initial encounter with biology was not at an institution renowned for pedagogy. I had been thrown out of UCLA after my freshman year for rabble rousing and other "subversive acts" (this was during the witch-hunting Joe McCarthy years of the late forties and early fifties), and had had to transfer to Los Angeles City College. Biology there was represented by a large lady of advanced years whose priorities were straight lines, neat penmanship, and appropriate use of stippling, the shading of drawings being forbidden. The course consisted of tedious dissections of earthworms, crickets, and other hapless insects that had been drowned in formalin. It was made evident that "good" biologists were those whose artistry allowed for neat depiction of the subjects' organs; other possibilities, if suggested, were summarily dismissed as impertinent. Extreme though this course may have been, it differed only in degree from many other introductory courses. Indeed, it is now no surprise to me that few, if any, of the older prominent figures in ethology came up through the academic system. Danny Lehrman, who became one of the world's foremost ethologists, confessed to me that he was a perpetual truant, who never would have gone to college except that he became a passionate birdwatcher and found his way to the American Museum of Natural History. There he encountered

T. C. Schneirla, the famous specialist on ant behavior and curator, who befriended him.

In my case, the disillusionment and despair (I simply couldn't manage elegant drawings, and my penmanship has driven my assistants to the edge of madness) was mitigated by earlier experiences as well as by the second semester, which quickly followed.

Semester two was taught by a hyperactive developmental fanatic known as Doc Bell. Embryology was his passion, and while I don't know whether he ever made original contributions to the field, I do know it was not for lack of enthusiasm. When describing the movements of cells after gastrulation, he would act the part of the embryo, tearing off coat and jacket in a passionate striptease that reenacted the movements of tissue layers. Dr. Bell was also addicted to Bach's sonatas and quality tea, both of which saturated the atmosphere of his lab. The relationship between aesthetics, emotions, and science could hardly have been better symbolized. Alas, midway through the semester Dr. Bell collapsed with a coronary, his place taken by a colorless if affable young instructor. It was not until my senior year, back at UCLA, that I realized that Bell, though unusual, was not unique in his passions.

My guess is that most biologists (at least those who, like ethologists, study organisms more or less holistically) have a history of exposure to woods and fields. The negative influence of so many introductory courses could otherwise scarcely be overcome. In my case, for example, I had a father with a devoted interest in horticulture. His prized possession was a three-volume set of Luther Burbank's original work, containing some of the first color photos to be printed. My father died when I was seven, and I received these books as solace. I believe I'd memorized them before I was eight. (Later, I was much disappointed to learn that botanists generally held Burbank in low esteem, and, indeed, his approach to breeding studies would not meet modern criteria for good experiments.)

Others in my childhood must have furthered my view that science was fun, for my earliest recollections are that I was going to study "nature." Curt Lewin, one of the founders of gestalt psychology, was a friend of the family. His children, with whom I spent a happy time after my father's death, had a myriad of pets. I particularly remember a pet rabbit that seemed to me to be quite unrabbit-like in its demonstrations of affection. I wondered then if this was due to its early rearing.

One of my uncles, Bruno Klopfer, introduced the Rorschach technique of projective testing to U.S. psychologists. From the age of about 5, and annually thereafter, I was a demonstration subject at his Rorschach workshops, first at Columbia University and then, after he moved to UCLA, on the West Coast. As I grew older, I was permitted to sit in on the evaluations and discussions (except, of course, of my own responses, whose changes were being longitudinally charted). Many of the participants were Jungian disciples and tended to invest early experiences (including prenatal ones) with a particular importance. I recall meeting the social psychologist Otto Klineberg and being impressed with the fact that so wise and important a man could take seriously my speculations about the Lewins' rabbit and possibility of rearing conditions having altered its behavior from the norm. All of this, in any case, must have insulated me from the unsavory character of beginning biology, and I suspect many biologists would tell a similar tale.

Another early experience equally biased my later behavior. During the three or so years of my father's fatal bout with Hodgkin's Disease, I became a member of the Knight family. Alice and Thomas Knight were Elders of the Abbingdon Friends (Quaker) Meeting, and close friends of my father in particular. They were themselves childless, but happy to adopt waifs from homes that were in troubled circumstances. Our home was. My mother was devastated upon learning that her young husband (he was 37 when he died) had an incurable disease. She had no marketable skills, the family fortune could not be removed from Germany, and she had to cope with the demands of two always-feuding boys, aged 2 and 4 (I was the older) along with two troubled (though not troublesome) German refugee children whom she had just taken in. We children were dispersed to different temporary homes and the invitation by the Knights that I should live with them turned out to be a godsend.

The Knights had a minifarm in Wyncote, in the early thirties a bucolic refuge off the Main Line to Philadelphia. Trees, birds, and a large garden and orchard made this a private paradise for me. The Knights belonged to the traditional Hicksite wing of the then-factionalized Society of Friends. In the early part of the nineteenth century, Elias Hicks and his followers, largely rural Friends of a tolerant, liberal persuasion, split from the urban Meetings. (Some have accused them of not even being Christians: it was claimed they recognized all persons as sons—or daughters—of God, and admitted to the legitimacy of prophets other than Christ.) The division was not to be healed until the middle of the next century. In the interim, Friends such as the Knights hewed not only to the Inner Light, that beacon of Quakerism, but also the outmoded traditions: no pictures, except for photos of people, adorned their walls; no music; food was to nourish the body, not tickle the palate; and simple speech was preserved. In later years, when I wrote Aunt Alice, she regularly reproved me for my too-flowery style— especially when, as an adolescent, I went through an Edgar Allen Poe phase.

Simple speech also meant using "Thee" and "Thou" within the family, a curious irony. When Friends of the seventeenth century dropped the formal "you" for the personal "Thee," it was to protest the class and rank differences that the personal pronouns marked. Nineteenth-century Quakers then introduced a distinction between Friends and others. I was told of a Quaker child being sent to a Jewish butcher in Philadelphia, who, intending to be cordial, gave her the purchased meat saying "Here thee is." "You can't say that," responded the child, "that's our language."

The Knights were far from obsessed with trivial traditions, however: these were merely the outward evidence of their deep commitment to a life of service. I spent hours with Aunt Alice before her wood-burning stove preparing and canning food that Uncle Thomas and I would then deliver to families in need. We also spent many hours each week in meetings where the problems of the world were reviewed and individual responsibility for each assigned, always in the context of worshipful meditation. As a child, I was ignorant of most of the specifics, but I am sure all of us children of the Meeting were impressed by that aura of personal responsibility. From an early age we clearly knew the significance of the phrase, "It is better to light a candle than curse the darkness."

The foregoing may explain why I was deemed too subversive for UCLA. My principal sin had been to initiate a campaign to remove the requirement for military training (ROTC) and the obligatory loyalty oath demanded of entering males. As I had originally enrolled as a freshman, I would have normally been subject to this requirement even as an upperclassman, but the university registrar persuaded the committee on military affairs to exempt me and allow me to return for my junior year. He personally favored compulsory military training as a way to "teach discipline," but evidently felt I was sufficiently "straight" to be spared further harassment. This was a fortunate circumstance, since the U.S. Supreme Court had some years earlier upheld the constitutionality of the training requirement, even for those who met the standards (in those years very narrow) as conscientious objectors.

The advanced courses in zoology at UCLA were a joy, though most of my classmates were premeds and then, as now, more concerned with grades than substance. This resulted in the minority who were truly interested in biology being favored, a not inconsequential factor in a large institution.

My introduction to embryology was through a course taught by one H. Schechtman, who lectured facing the blackboard, drawing his way across the board with his right hand while erasing the drawings with his left. In this way he could repeat his circuit without interruption. As he was also hard of hearing, questions were of no avail in any case. The only way for students to record both lecture and sketches was to pair off and divide the chore. We learned to bring carbon paper to that class.

Schechtman's detailed expositions are a distant memory, but the message he delivered was certainly not lost. I'm sure my view of organisms as dynamic systems that are ever changing as a consequence of both prior and present influences was formed by Schechtman. As a result, I continue to believe that training in embryology is an invaluable asset to an ethologist.

George Bartholomew offered a fascinating course in physical ecology, a study of the manner in which organisms adapt to the demands of their environment. It was my first introduction to the subject. The personal but unassuming manner in which he described some of his own work gave an immediacy that I always wished to be able to emulate. Once he told the tale of falling from a ladder while collecting sparrows at night from amidst the ivy-covered walls of a Harvard hall. He was unaware of having sustained any injury until the next day, when his students led him from the classroom to the infirmary. During his presentation he had apparently been unconsciously repeating sentences three and four times as the result of the concussion he sustained. Thereafter in our course, each time he repeated a comment, whether for emphasis or in error, he would throw up his hands and assure the class he'd not been on ladders the previous night.

While Bartholomew taught physical ecology, he was equally interested in animal behavior, and particularly in the interface we know today as behavioral ecology. He focused on thermal regulation and its role in behavior, but his wide-ranging studies covered a multitude of other topics as well. His lively descriptions of the work of his friend Don Griffin, who analyzed the manner in which bats could exploit lightless environments, inspired numerous students long before Griffin's own writings on echolocation had become widely known. A tutorial semester I spent with Bartholomew further confirmed me in the profession of biology and particularly animal behavior. I often encounter other biologists who had comparable experiences with Bartholomew. Biologists such as he, prolific yet modest about their own work, always supportive and helpful, are like the nodes of computer networks and have disproportionately great influence on recruitment to their field.

Yale in the early fifties was still the home of R. M. Yerkes, the famed psychologist/primatologist, and a program in psychobiology, to which Bartholomew urged me to apply. I eventually followed his advice, though only after a year's stormy hiatus, the indirect result of the Korean War and Truman's ideas on universal military training.

As a Quaker I was unable to accept conscription. I had registered as a conscientious objector, though not without misgivings, and then subsequently returned my draft card. I was promptly arrested and sentenced to serve three years in prison. Due to the pleas of several of my teachers, including particularly Bartholomew and a young political science professor, the judge postponed the start of my sentence until the day after commencement. This was a mixed blessing, because I'd ceased attending class since my arrest in February. With class notes from friends and after a number of sleepless nights, I did pass my exams. On the day my sentence was to begin, the judge suffered a disabling heart attack and his place was taken by Leon Yankwich, chief justice of the Southern District, and a leftover from Rooseveltian days. He proceeded to resentence most of his subordinates' victims, releasing me on probation on condition that I "obey all laws, State and Federal, so far as . . . conscience allows." This unusual condition assured my survival without a draft card. Thus protected, I joined a technical assistance training program (forerunner of the Peace Corps) at Haverford College, which was also the home of the herpetologist Emmet Dunn.

Dunn's fame rested on his biogeographical studies—the distribution and abundance of species —but his broad interests encompassed snake behavior as well. He was particularly curious about the reasons for regional differences in the food preferences of particular snakes, differences that work now suggests are due to early experience with particular prey. He was also intrigued by the "coral snake mimic problem," which he assigned to me when, bored by endless seminars on technical assistance, I'd knocked on his door.

There are numerous snakes whose banding patterns resemble that of the poisonous coral snakes of the genus *Micrurus*. Resemblance between sympatric species (those having overlapping territories) were assumed to reflect instances of Batesian mimicry, but it was not clear how these came to be established. (In Batesian mimicry, an unpalatable or dangerous "model" is imitated by a palatable or harmless mimic.) However, if a coral snake (the model) succeeds in biting a potential predator, the latter is likely to die, and not being in a position to profit from its experience, it is unlikely to be much impressed by the colors of harmless mimics. Or was there another explanation?

In fact, there proved to be several. The most persuasive stems from the determination that many of the mimics are at least mildly poisonous, so we could be dealing with Müllerian, not Batesian, mimicry. This is the situation in which all the species involved are unpalatable. By resembling one another they simplify the task of recognition and avoidance by their potential predators, who are therefore more likely to spare them. Further, we now are more willing to believe that some patterns are dictated by morphologic constraints and are not necessarily adaptive. While I started thinking about this problem at Haverford with Dunn, the first explanation to be tested was one suggested by Jane Van Zandt Brower. Her proposal, made several years later, after I'd moved on to Yale, was that "friends" of the potential coral snake predator, seeing its death throes, would, by observation, learn to avoid similarly marked snakes. My first experimental results were the demonstration that this could indeed happen. I used domestic mallards as subjects and my "snake" was a paper pattern surrounded by an electric grid that produced a powerful foot shock. Corn on the pattern attracted a duck, but only once. Its frenzied response resulted in none of its companions even coming near the corn. Even ducks that merely witnessed this avoidance, but not the actual shock response, eschewed the pattern. As I was still young and inexperienced, I couldn't believe this was anything but a highly original observation that deserved a special status. Thus, I dubbed it "empathic learning," although the perfectly respectable term "observational learning" was already in the literature (which I'd not read). To my later embarrassment, the results were published with the unfortunate title: "Empathic learning in ducks." Not surprisingly, it was as often cited as "Emphatic learning . . ."

Dunn insisted on a regular review of the scientific literature, rather than the eclectic habits I'd developed at UCLA. The Haverford library was not particularly distinguished, at least not insofar as scientific journals were concerned, but it did contain the *Quarterly Review of Biology*, where I encountered Niko Tinbergen's treatment of displays. This article, more than any other single paper, served as a "releasor" for my ethological inclinations.

"Derived Activities: Their Causation, Biological Significance, Origin, and Emancipation during Evolution" was the imposing title of Tinbergen's 1952 *QRB* paper. In it he sketched the functions and origins of instincts as developed in his earlier book, *The Study of Instinct* (1951), and work by Lorenz. Then, after describing the predictable course of innate acts, he went on to remark on the "odd" circumstance that seemingly irrelevant acts are sometimes interposed. Two skylarks "engaged in furious combat, suddenly peck at the ground as if they were feeding; it is equally surprising when starlings during territorial fight begin to preen their feathers, only to resume their fight a few seconds afterwards" (p. 6). Such behavior, often observed but unremarked upon, was what he termed "displacement"; that is, a set of behaviors were exhibited out of their ordinary context and needed to be accounted for in some way. He proceeded to explore the origins and significance of these kinds of behavior in terms of evolution.

In this landmark paper Tinbergen literally opened a door to a new domain of inquiry, one that is still being explored. Its importance was evident to many naturalists and fledgling ethologists from the start, but, more important for the novitiate, it led one step by step through the evolutionary arguments on which ethology rested: the explanation of the origins of communicative signals. Since no texts or other comparable systematic treatment existed at that time, I suspect Tinbergen's 1952 paper probably served as an initiatory experience for all would-be ethologists of my generation. Indeed, for me, it was the crucial factor in resolving that ethology was the subject for me. (A further minor factor was probably that I slaved so hard to master some of Tinbergen's erudite vocabulary—such delicious words as "autochthonous"—that I would have resisted having to abandon them.)

Another piece of scholarship that I came upon at this time was the early work of Ramsay and Hess. Eckhard Hess, an erudite and urbane professor of psychology at the University of Chicago, was among the first to approach the study of imprinting experimentally.* (Actually the Swede, Eric Fabricius probably has priority, but Fabricius's modesty makes it impossible to be certain. Hess, in all events, was first to bring the phenomenon fully into the laboratory and to study it using all the automated gadgetry so dear to psychologists of the Skinnerian generation.) What endeared Hess to ethologists—who would have disdained a purely laboratory approach—was that he was equally committed to field observations. These he undertook at his vacation retreat in Maryland, where he found a congenial collaborator at a local high school (the McDonogh School) in A. O. Ramsay.

The work of Hess and Ramsay was solid, but, more important to me at the time, it suggested the possibility of remaining active as a researcher even if only high school science posts were available for support. In the early fifties, the prospects of actually earning a living as a college professor were not bright for an aspiring ethologist, particularly one with a prison record and a disinclination for the loyalty oaths required at some public institutions. Ramsay was encouraging, and Hess, whose lab I visited in 1954, all but coercive on the issue of the practicality of such a mating. Hess also demonstrated that his most interesting work did not, in fact, require the apparatus of his well-equipped lab. His finding that pupillary dilation was a good measure of affect in humans, for instance, required only a split prism and a simple camera. With this device he ultimately earned considerable money previewing *Playboy* magazine's "Playmate" photos. The degree of pupillary dilation of his male subjects became the criterion of selection.

And so, an unexpected phone call offering me the post of science teacher

---

* Both were anticipated by Spalding in the nineteenth century, but this work went unnoticed until discovered by Haldane and Spurway in the 1950s.

at my former high school, an international boarding school in Lenox, Massachusetts, was the catalyst for a move from Haverford's collegial atmosphere to the austere beauty of the Berkshires.

Ramsay had set a grand example, combining good science with high school teaching. The year in Lenox, however, convinced me that while this was a viable option, I still had more to learn first. I finally did as Bartholomew had urged, and applied to Yale. In the interim, however, I had resumed a friendship with Martha Smith, whom I met at an American Friends Service Committee work camp while I was still at UCLA. At that time, we each had separate boy-/girlfriends, but found reasons to spend a fair bit of time together. Our mutual problems with our "others" soon were resolved by our shedding them, and, after a year of cross-country corresponding, becoming engaged. Martha transferred from Pomona to Mt. Holyoke College, and for another year we endured the train ride between South Hadley and New Haven, before we were wed at her parents' home in the summer of 1955.

**PK in early days (1952, actually), while teaching school in Lenox, Mass.**

# 2

## Yale Years

### OSBORNE'S INHABITANTS

It was principally Yerke's psychobiology program that had led Bartholomew to laud Yale as a Mecca for students of behavior. That program came to an end shortly before my arrival in New Haven. However, no one meeting zoology's director of graduate studies, G. Evelyn Hutchinson, could be expected to withstand the seductive charm of his manner. There are, unfortunately, not many learned scientists who have the capacity for evoking the very best in their students, without stridency and certainly without being patronizing. The first few minutes of our conversation convinced me that this was the mentor I sought, even though his own behavioral work had been restricted to an obscure study of sadomasochism in snails (Hutchinson 1930).

I actually almost missed meeting Hutchinson. When I arrived at the Osborne Lab, already behind schedule on my rounds of the faculty, Hutchinson was occupied with one visitor and another was in the hall awaiting his turn. This last man was about to turn in his bound thesis and, as we engaged in conversation, displayed his results to me: utterly cryptic vertical graphs that he said represented changes in pollen abundance through time. Neither the methods or purpose of pollen stratigraphy were known to me then, and Dan Livingstone was disbelieving that anyone coming to Yale (a center for such work, under the leadership of Hutchinson and another Renaissance man, Ed Deevey) could be as ignorant as I actually was. I felt hopelessly out of place as Dan outlined his conclusions to me, and after he entered Hutchinson's office I nearly fled. Fortunately, he emerged before I'd totally lost my nerve, and that proved the start of a friendship that still endures.

That first meeting with Hutch, whatever else it provided, also led to the only bit of bad advice I received from him. After delivering a lengthy monologue on how I hoped to compare song learning in passerine and nonpasserine

27

birds (perching birds and their predecessors) as a clue to behavioral evolution, Hutchinson suggested that I familiarize myself with the writings of Freud. What work in particular?, I wondered. "All of it," he directed. I dutifully went through the *Gesammelte Werke*, and to this day am convinced that there were innumerable better ways to have spent the many hours this cost me.

**G. Evelyn Hutchinson**

A more profitable suggestion from Hutchinson was that I cross town to the offices of Frank Beach, the doyen of reproductive behavior. The redbrick of the medical center, where the psychology department was located, contrasted unfavorably in my eyes with the aging stone and ivy of the central campus, including the Osborne Lab. The contrast to the cluttered dignity of Hutchinson's rooms was even greater. "Come on in!," someone bellowed in response to my timid knock. The room I entered was bright and sun-filled, allowing me to view walls cluttered with photographs and cartoons—all devoted to the depiction of genitalia belonging to everything from alligators to zebras, in various stages of distention. My host, in dirty khakis and a stained T-shirt, had his feet on the desk, a can of Schlitz in his hand. "Have a beer, sit down, tell me what you're doing," was the staccato command. As I timidly began, he interrupted. "We better get Julian in here," and he reached for the phone.

Julian Jaynes, now at Princeton and author of a much-discussed book, *The Origins of Consciousness and the Breakdown of the Bicameral Mind* (1976) was then Beach's graduate student investigating aspects of imprinting behavior in fowl (the advantage of which, he pointed out, was that they tasted better barbecued than Beach's usual rodent subjects). Jaynes was a philosophically inclined experimentalist, whose diffident and cautious manner contrasted with Beach's ebullience. Over coffee at a nearby beanery we argued about imprinting, its prevalence, significance, and evolutionary development, a conversation that revealed Beach to be as profoundly interested in abstract issues as the decor in his space seemed to deny. The patina of vulgarity he displayed was misread by many who did not know him well. Even Hutchinson, in reviewing one of Beach's many books, wrote that while measuring the human penis in metric units might be science, using English units was obscenity. Beach favored inches.

Hard drinking and jocular though Beach was, those who conversed with him were soon disabused of their prejudices. Beach's book *Hormones and Behavior* was a seminal work that related physiology and behavior. His analytic intellect also served to clarify many of the vague propositions and contradictions of nativistic theorists, especially Lorenz. His "The Snark was a Boojum" (1950) is as erudite an analysis of the problems with instinct theories as one could wish. In this essay he demolished the notion of an instinct-learning dichotomy in as entertaining and edifying a manner as is possible. More than any other comparative psychologist of the past half century, Beach has revealed the importance of comparative studies if one wishes to understand the evolution, development, and mechanisms of behavior.

Sadly for me, Beach was away from New Haven much of my time there, so our contacts were fewer than I would have liked. There's no question, however, that many there, myself included, profited immeasurably from the careful and thoughtful review that Beach gave his students' work. About my only difference with him in the years that have passed is in his choice of beers.

The Osborne Labs at Yale are not known as a cradle of ethology, although an extraordinary amount of the work done there was as a tributary to the swelling ethological river. There was Talbot Waterman, for instance. He was not popular among graduate students, who were sternly admonished by him for their lazy ways whenever caught browsing a newspaper (rather than *Science*). Waterman had discovered how *Limulus,* the horseshoe crab, detects polarized light and determines its plane of polarization,

**Yale's Osborne Zoology Labs**

enabling it to orient and navigate. The method was simplicity itself: An oscilloscope recorded the output from electrodes implanted in the optic nerve. A polarizing lens was placed between the eye and a source of unpolarized light. The activity of the nerve waxed and waned, reaching minimum or maximum values for every 90° whenever either the eye or the lens was rotated. It was an elegantly convincing demonstration, for which Waterman was justly admired, and it contributed greatly to the design of studies on behavioral mechanisms.

Charles Remington's interests were in the biogeography and systematics of lepidoptera (moths and butterflies), but he encouraged his student Jane van Zandt Brower to examine the behavioral basis of mimicry in the viceroy-monarch butterfly complex (which had led me to my "Empathic Learning in Ducks" study). Indirectly, at least, Remington helped promote this work, for he was genuinely interested in encouraging his students to find their own paths. Indeed, this was the outstanding attribute of most of the Yale zoologists.

Chair of the department was John Sterling Nicholas, a stern and forbidding figure to graduate students. Osborne had a dress code for its graduate students: if not a lab coat, then tie and jacket were required. In the fall and spring Robert MacArthur, a fellow grad student, and I would go birding for an hour or so early each morning in nearby West Rock Park. We stocked clean clothes in the lab, but obviously had first to enter Osborne's hallowed halls in our muddy field gear. It was never pleasant to meet Nicholas under those circumstances. His look if not his words made his lack of appreciation for the demands of field work clear. Nonetheless, he was as committed to the comparative method in biological studies as any biologist I've met. His lectures on the principles of homology (similarities in structures as a consequence of common ancestry), though I later came to disagree with them, certainly helped prepare me (and others) for the emphasis on behavioral homologies as the clue to evolution.

Nicholas's sixtieth birthday was the occasion for one of the graduate students' more successful (and less harmful) pranks. The faculty had prepared an album of photos of Osborne and its inhabitants. Most had been taken by Waterman, an excellent photographer. We discovered his negatives in the darkroom, however, and noted carefully the scenes he'd recorded. We then restaged the scenes, with certain minor changes, and photographed them for an album of our own. Waterman's collection showed one picture of the departmental tea, a smiling teaching assistant handing a cup to a professor. Our version had a hand reaching out from beneath the tea-cart grasping a bottle of poison from which a few drops spilled into the tea. A photo of a seminar group showed the table surrounded by earnest faces, some supported by hands, one or two by bare feet. A portrait of the arcade by Osborne's entrance originally showed a display case with various primate skeletons—from monkey through ape to man. The last we replaced with one of our own, skinny Joe Shapiro.

Waterman presented the faculty's album to Nicholas amidst considerable pomp and circumstance. The gift was accompanied by a "speech" in Latin, which impressed almost everyone. Hutchinson and I, who were standing together, were convulsed by laughter. "Exceptio probat regular de rebus non exceptis ars gratia. Artis audemus. Juro nostra defendere de gustibus non disputandum est," and so on. The intonation was good enough to obscure the meaning to all but Latin buffs, of whom, apparently, few were present, so no one else realized the speech was but a string of unrelated clichés.

We handed over our copycat album rather surreptitiously. It was much appreciated by Nicholas, to the evident and outspoken annoyance of the faculty.

J. P. Trinkaus was one of the few who appreciated our joke. Indeed, his youthful appetites endeared him to most of the students, even those for whom experimental embryology was not a favorite subject. Since it was one of mine (and has so remained), I was especially attracted to Trink. His commitment to the American Civil Liberties Union, in which I was also active, provided yet another bond.

Trinkaus's main project at this time involved the issue of whether, when embryonic tissues were disaggregated and then allowed to recombine and regenerate, their cells dedifferentiated and redifferentiated; alternatively, did they maintain their original identity and migrate to the appropriate place? For example, embryonic chick mesonephros and cartilage form a specific spatial pattern. If the cells are separated (with trypsin) and mixed, then allowed to continue development, the original pattern is soon restored. Did the cells migrate back to their original places? Or did cells in the place for cartilage develop into cartilage cells, even if they'd been mesonephros? I've forgotten which it turned out to be, but I found it a fascinating question.

Dillon Ripley, ornithologist and the patrician polyglot head of the Peabody Museum (later, as secretary of the Smithsonian Institution, he was responsible for major reforms in that institution's activities and outreach) was another major resource to the few of us at Osborne interested in behavior. Aside from the seemingly limitless funds with which he supported his students (some of which, I suspect, were taken from his own pocket), he had a private waterfowl preserve in nearby Litchfield, Connecticut. This consisted of an irregularly shaped pond, landscaped so as to seem far larger than the several acres it encompassed. It and the surrounding meadows and varied woods served as home for a collection of waterfowl from the far corners of the globe. Wallabies and muntjac (or barking deer) also wandered about.

Ripley's preserve was so laid out that substantial numbers of the different species of ducks and geese could coexist and breed. Thus, they provided an unparalleled opportunity for comparative behavioral studies, unrestricted by the kind of rules that zoos and public institutions must impose on researchers.

Ripley was unstintingly generous in making the preserve available to his students. Since the birds had to be provisioned, especially during the breeding season, he hired me as a supernumerary caretaker. My task, in addition to feeding, was to collect eggs for artificial incubation. This provided me with hatchlings for laboratory experiments and also resulted in the hens producing replacement eggs, increasing the total output. This system did have some drawbacks. The black-necked swans nested on an island that I could only reach by donning waders. The ungainly garb made me no match for the alert, maneuverable, and large swans who defended their nest against my marauding with a vigor that often defeated my nerve.

Another problem lay in the fact that we originally had but one old-fashioned incubator available. It was set up in my room at Osborne; limnologist Joe Shapiro was next door, together with his elaborate vacuum distillery. (Along with creating our party fare, he also used it to obtain residues from his limnological samples.) Joe blew a fuse late one night, and not realizing that we were on the same circuit left it to the morning to replace. My old Sears incubator lacked a no-power alarm, so most of a season's worth of eggs was lost that night. Thereafter, the eggs were distributed between incubators tied to three different circuits.

Ripley's preserve wasn't the only basis for support he provided his students. As head of the Peabody Museum he also could hire help for the many chores there. When my classmate MacArthur and I were particularly low on cash, he paid us to go through the mammal collection and reorganize it, a job we'd happily have done for nothing. The collection was in a mess, though, with many labels missing or incomplete. The original logs were accurate, however, so in most cases we were able to identify things. One twisted piece of fur defied our best efforts at identification, so we casually tossed it into the refuse pile. We were mortified when, the next morning, Ripley presented us with the skin. "Imagine," he said tactfully, "this [priceless] exemplar of [Wojabbus] was in the rubbish pile this morning. I can't imagine how it got there. I believe it's number 4897 in the acquisitions log." The specimen, untagged and, to us, unrecognizable even as to family, was properly listed.

A good part of our discard pile did pass muster: it consisted of skins of various breeds of local dogs with labels such as "Mrs. Smith's Rex," or "unknown, killed by car on York street." The assemblage was evidently the work of an avid collector and predecessor of Ripley.

Unfortunately, Dillon Ripley, a courtly gentleman with impeccable manners, refined wit, and extraordinary linguistic gifts, did not get along well with Frank Beach. The latter, though no intellectual inferior, seemed to revel in the image of "ordinary Joe," uncouth, roughly clad, and disdainful of societal niceties. Their mutual aversion was sufficiently intense that it was less than pleasant to meet with them together, which posed a problem for me since both were members of my doctoral committee, and neither would agree to the changes in my thesis the other wanted. Hutchinson provided the solution by having me prepare separate and (slightly) different versions for each to review and sign.

Professor Alex Petrunkewitch was nominally emeritus by the time I reached Yale, but his zeal and activity belied his status. The weekly tea with Professor Pete was a major part of the Osborne routine and the venue for discussion or argument of all the latest controversies. Professor Pete also lectured frequently, generally on some aspect of arthropod phylogeny. He was a dedicated proponent of saltatory evolution, the view that the morphological differences between supraspecific taxa are less often due to the accumulation of small mutational differences as to single-generational changes in chromosomal organization (macromutations).

Pete's other important claim was that, at least in spiders (his special group), behavioral attributes were as conservative as morphology and thus could be used for taxonomic purposes. In this he antedated O. Heinroth, who had made a similar argument for waterfowl, providing thereby the impetus for much of the work of Konrad Lorenz. I do not know, however, whether Heinroth was familiar with Petrunkewitch's work; Lorenz claimed not to have known of it when I once asked him. It certainly influenced me.

As an "emeritus," Petrunkewitch was no longer able to have students. He groused that he was just now (in his seventies) hitting his stride—this was the time, he argued, when students would benefit most from him. I was not inclined to doubt him. One of his remaining responsibilities was to administer the foreign-language exams. Yale zoologists were expected to have a reading knowledge of two foreign languages relevant to their field of inquiry. For me these were German and French, both of which I'd studied in high school. I took a refresher course in German, and passed that exam without trouble, but I lacked money for a French course, too, and also the discipline to study properly on my own. Professor Pete's examination *modus* was to select a book in the student's field (and language) and have a selected page read aloud and translated. Knowing my interest in embryology, he selected a late-nineteenth-century French classic, opened it, and passed it to me. "Ask me any words you need," he offered. Single words weren't going to help me, however. The lengthy Gallic expositions on embryonic events were altogether beyond my ken. Professor Pete mistook my silence and desperate fixation on the page for fascination with the subject. "An interesting idea, you think?" he interjected. "Professor Pete," I intoned in my most serious voice, "do you think he [the author] really believed this, or was it a pose?" I don't recall the details of Professor Pete's response, but it was a lengthy, involved exegesis of the section with which I was struggling. The few sentences I could translate finally found a matrix into which they could be embedded.

Edward Deevey, a limnologist professionally but a veritable Renaissance man, was among Hutchinson's most brilliant protégés and, like his mentor, kept abreast of developments in fields far outside of his own. His position as editor of *Ecology* no doubt helped. It was in this role that he unwittingly helped initiate a twenty-year research program on mother-infant attachments in goats. It resulted from Deevey handing me page proofs of an article on socialization in goats by Nicholas Collias. Collias was eminent as a student of bird behavior, with an emphasis on social organization. However, for one season he had allied himself with the psychologist Howard Liddell, at Cornell, for the study of attachment behavior in goats. Their work suggested that the bonding between kid and doe depended on events that could only transpire in the first few minutes after parturition. I had been fond of goats since a summer in childhood spent at a goat ranch in the Mojave. Liddell was trained in the Pavlovian tradition and was no naturalist. Collias was a good naturalist but his strong suit was ornithology. Despite my being woefully less prepared for scientific work than they, these differences in our backgrounds convinced me they had not described all there was to be seen in the interactions of a doe with her newborn. I was further struck by the similarity between Collias's descriptions of bonding between kid and doe and Lorenz's of imprinting of goslings on their parents. Were these "homologous" phenomena? If so, what a wonderful system for tracing the evolution of "imprinting" behavior.

Correspondence with Liddell revealed him to be a friendly and helpful gentleman, eager to take on and assist students. Recognition of his work had been late in coming. When I met him, shortly before his retirement, he had just received what he said was his first invitation to chair a major meeting. He was much touched by the honor he felt this did him, and displayed not the slightest trace of irritation at having been so long neglected. Collias, incidentally, similarly impressed me by his unassuming manner; he seemed altogether free from any need for recognition, a trait I lamentably often found lacking among other academics.

Liddell invited me to come to his lab for postdoctoral work, which I immediately resolved to do. This did not in fact come about, though all arrangements had been made. Nonetheless, we remained in touch for several years and the small herd of goats I built up while in graduate school went to Liddell when I went on to Cambridge.

Goats were admirable subjects for studies of mother-young relations, especially by impecunious graduate students. They could be bred in their first year, after which they paid for themselves (milk, butter, cheese, all of which we produced). They generally bear twins or triplets (once we had quintuplets: all survived), so every subject had a control. They were agreeable, social animals that even very young children could manage (our first daughter did the milking by age five); the excess kids could be eaten (probably much of the "lamb" one bought in those days was really "kid"); the bucks, which could be pungent in the fall, could be replaced with frozen semen; in short, the goat was a veritable schmoo.* Indeed, were it not for

---

* Al Capp's famous animal, which lived only to be able to satisfy the lusts of its owner.

the schmoolike attributes of goats, Martha and I might never have ventured into a subsistence farming enterprise that for many years sustained our family of three daughters and the friends who came to live with us.

The buck-smell problem, unfortunately, was not taken seriously enough by me. For years, whenever I entered a room, colleagues would turn their backs, sniff ostentatiously, and wave "Hi Peter" over their shoulders.

My thesis research had to do with imprinting in waterfowl, so the goat work had to be an ancillary, unsupported project. As a consequence, I began it modestly, with a single pair of newly weaned kids. By the time I completed my graduate years, I had done experiments with less than a dozen does, but the results, which confirmed Collias's findings that the first five or so minutes after parturition were crucial to bonding, indicated this was indeed a promising model. Not that there weren't problems. Foremost was the need to be present at the moment of birth, which resulted in countless sleepless nights. On more than one occasion, birth occurred precisely during the time I took a toilet break.

One goat in particular is etched in memory. She was due to give birth in January, a month that, in New Haven, is as bleak and cold as it gets. Martha and I were sharing a lovely farmhouse beside the woods of Bethany with Robert and Betsy MacArthur and their year-old son, Duncan. The farm belonged to the microbiologist, David Bonner, who'd rented to us while he took a sabbatical in France. There were sufficient outbuildings for our various animals, including the goats, Thorsten and Heidi II, but these huts were unheated, cheerless hovels. After several miserable nights keeping company with my goats rather than a warm, if less furry, wife, I was about to abandon the project. I'm not sure anymore whose idea it was, but suddenly the guest room was emptied of furniture; a canvas sheet was placed over the floor and bales of hay heaped against the wall. Into this warm sanctuary came Heidi. She continued her procrastination, however, and grew daily bigger.

One morning I could no longer delay going to the lab. Martha and Betsy were to take charge. It promised to be a full day for them. Duncan had gone berserk that morning. Always active, he had surpassed himself. Every piece of movable furniture had been rearranged and 90 percent of the books from the lower shelves of David Bonner's ample library lay on the floor. Bonner, the owner, and his wife, a woman with an unparalleled (in my experience) passion for cleanliness and order, would not have been pleased, and we guiltily acknowledged this.

As Robert and I left, waving adieu to our families, Heidi thrust her head from the open window of her guest room. "Baaa," she bawled, a fitting caption to the picture of our disordered household. We called Martha and

Betsy outside for a view and collapsed into each other's arms with laughter. "Thank goodness the Bonners are in Paris," I said, as Robert and I drove out the lengthy, bumpy driveway.

Just before we reached the road, an unfamiliar car turned in. We squeezed to the right to make room, and stopped as an all-too familiar figure leaned out of the driver's window: "Hi fellows, I flew in from Paris for the weekend. Just thought I'd pick up a few books and things," said Professor Bonner. We mumbled something about an early exam, sped into New Haven, and hid in a friend's lab all day.

On returning home to strangely silent wives, we asked no questions and were offered no account. Never, not to this day.

The only echo of this trauma came years later, when I met Richard Setlow, a Yale biophysicist. He had come to Duke to give a zoology department seminar. When we were introduced he thought for a moment, then burst out with a grin. "Oh yes, you're the one who had Bonner's house." Then he exploded with unstoppable laughter. I fled.

An ethological education at Osborne was as much dependent on fellow students and visitors to the university as it was upon the Yale faculty. Graduate students were not expected to fill their days with lectures and formal seminars, but to initiate research soon after their matriculation. Thus, informal student-organized seminars and discussions played an important role in our education. In the latter half of the fifties one student in particular left a mark on his peers. This was my friend, Robert MacArthur.

MacArthur had been an avid naturalist from early childhood. His father was well established in mammalian genetics, while his mother taught bacteriology. Perhaps it was adolescent yearnings for independence that led him to mathematics, but after earning a master's degree in the subject he elected to rejoin the family of biologists. Now, however, he had coupled his intimate knowledge of natural history (he could correctly identify any North American bird, beetle, or bush at a glance) with the skills and insights of a competent mathematician.

The hoary stereotype of the "rawboned, square-jawed" Vermonter certainly fit MacArthur when he first appeared at Osborne. He wore homespun fabrics, and this added to the backwoods impression. His quiet, understated articulations and slight social awkwardness also contributed to the stereotype. He and I identified commonalties almost immediately, and commenced a friendship that continued to grow until his death from cancer at an early age in 1973.

We were assigned to teach together in a laboratory for embryology and comparative anatomy, subjects dear to my heart and mind and totally alien to Robert's. In return for an enormous amount of tutoring in calculus and statistics (my weakness), I covered for his embryological inadequacies. It was a mutually satisfying collaboration, soon augmented by early morning birdwatching.

Robert's wife, Betsy, was still at Smith College during his first year, completing her thesis in botany, and only joined him on weekends. I was not yet married, so we two had no one else competing for shared time and ideas. (Later on, when we both were living with wives, we were delighted that Betsy and Martha also became close friends, with a great many shared interests.) Those interchanges opened my eyes to the close interdependence of behavior and ecological factors (not yet taken for granted in the 1950s) and Robert to the role behavioral studies could play in elucidating ecological puzzles.

Among the first issues upon which we began speculating was the still intriguing question of why the tropics seem to have a more diverse array of species than comparable areas to the north and south. The common view was that the tropics had a longer history of freedom from major climatic catastrophes and hence there was more time for the occurrence of speciation. Also, there were, presumably, fewer instances of widespread extinctions, as by glaciation. Robert had developed a simple measure of species diversity derived from the Shannon-Weaver expression for information content. It took into consideration the intuitively "obvious" fact that the number of species and their diversity were not identical. For instance, imagine two woods each with 100 individuals divided among 4 species. In one wood, however, there are 25 individuals of each species. In the other 97 individuals belong to 1 species and the remaining 3 species are represented by but a single individual. The latter woods is perceived as less diverse than the former though both contain the same number of individuals and species. MacArthur suggested that our perception of diversity was inversely related to the probability of being able to predict the species identity of a randomly encountered individual. The easier the prediction, the less diverse the woods. Thus, the Shannon-Weaver equation allowed a specific value to be attached to a particular level of diversity, opening the door to quantitative studies of diversity.

MacArthur's next step was to demonstrate that in the fields and forests of New England the diversity of bird species could be accurately predicted by knowing the diversity in foliage density at different heights above the ground. The greater the foliage-height diversity, the greater the bird-species diversity.

This relationship also seemed to hold true in the New World tropics, except that the slope of the line relating foliage-height to bird-species diversity was displaced upward for a given foliage-height diversity. Perhaps the "time for speciation" hypothesis was correct after all.

As we mused on this point, I suggested that temperate-zone birds likely had a more difficult life than their tropical cousins. Whether or not they migrated, their behavioral repertoire had to be sufficiently broad as to allow for the exploitation of different habitats and foods at the different seasons. The relative constancy of the tropics, I argued, might permit for a greater degree of stereotypy in behavior; perhaps tropical birds were, to a greater extent, "specialists" rather than "generalists" in their utilization of the habitat.

These thoughts did not arise from a vacuum, but had been triggered by work MacArthur had been doing with our mentor, G. E. Hutchinson, on the concept of "niche." The niche of a species, according to G. E. H.'s definition, was a unique hypervolume defined by all the factors that limited the survival of that species. MacArthur attempted to describe niches mathematically and to develop a model to explain differences in niche densities. The degree to which a species specialized in any domain, for example, in feeding, would correspondingly shrink its niche, allowing for more niches to be added. If climatic and general environmental stability were high, I reasoned, the resulting behavioral stereotypy would be equivalent to a reduction in niche volume. This might explain, in part, the greater diversity of the tropics. In all events, the subject commanded many hours of our attention (even when we were supposed to be preparing for our comprehensive exams), and ultimately led us to the tropics and publication of several articles. Our views finally turned out to be based on no small number of oversimplifications and errors, but, at the time, we were intoxicated with the thought that our ecology-behavior alliance had resolved a major issue.

Others of the Yale student community had interests that extended to ethology. One was Jane Van Zandt Brower, who, with her husband Lincoln, continued for many years to explore the behavioral and ecological puzzles posed by mimicry in butterflies.

There was also Alan Kohn, who became the leading expert on the snails of the genus *Conus*. While a graduate student, he discovered a species of *Conus* capable of poisoning its prey. His work on the several behavioral questions this raised continued for some decades at the University of Washington. Bill Cotter studied moths, and particularly genetic aspects of their behavior. He was several years ahead of us, so his conviction that genetic factors needed to be considered in behavioral studies was taken more seriously by us than might otherwise have been the case. It was a lesson whose implications took rather long to be absorbed, especially by comparative psychologists and ethologists in the United States (for whom a rat was a rat). Cotter had previously studied with E. Caspari, a pioneer in behavioral genetics, whose work dominated the field in its early days, but who is rarely acknowledged today. George Watson was a classicist who turned ornithologist in order to more easily have reason and (above all) money to visit Greece. (His change of field and travel were financed by funds from the ornithology collection of Yale's Peabody Museum.) George's interests and competencies in biology developed apace, and he more than any other of our colleagues focused us on the question of how subspecies formed and interacted, which, in turn, introduced us to the notion of sympatric and allopatric speciation.

The traditional view, emphasized in particular by Ernst Mayr, whom I met several years later, was that geographic separation was a virtual *sine qua non* for the differentiation of populations of one species into two or more. So long as contact and gene exchange was more or less continuous, the necessary genetic isolating mechanisms, he argued, would not be able to develop, with exceptions only among rapidly reproducing creatures such as microorganisms. The spontaneous increase in the number of chromosomes, as occurs in some plants, was another possible exception, but as changes in ploidy are rare among animals, this did not weaken Mayr's case.

## NOTABLE GUESTS

Osborne was a stopping-off place for a great many guests, and several of these brought with them visions of the new fields of ethology and behavioral ecology. There was, for instance, Frank McKinney, whose seminar on the breeding behavior of waterfowl introduced us to the classical Lorenzian approach. John King detailed descriptions of the social behavior of black-tailed prairie dogs and focused attention on the role of tradition in mammalian behavior. King's prairie dogs lived in large family groups, called coteries, whose communal territories were passed on from one generation to the next. The cause is the intense sociality of the animals, which spend a great deal of time in mutual caressing and play—particularly the young. The demands of the young apparently become onerous to older animals, which eventually abandon their territories and set up homesteads elsewhere, leaving the younger generation to carry on.

Neither McKinney nor King achieved great prominence, though their work has been often cited. McKinney worked under the ornithologist-art-

ist Al Hochbaum *(Travels and Traditions of Waterfowl)* at the Delta Waterfowl Research Station, located in the pothole country of Manitoba. The
station lay on the edge of the Manitoba prairie and was a center for waterfowl propagation. Every imaginable waterbird was to be found there and
the housing, incubator, and pen facilities of the station made this a Mecca
for students studying ducks. McKinney's great contribution to ethology
lay in part in his bringing Delta's existence to the attention of the scientific
world. Once he had described the facilities, I had no trouble visualizing
how ideal an environment it would be for my studies of imprinting, which
required large numbers of incubator hatchlings on a daily basis. Several
generations of my students followed in due course.

King, through his imaginative interpretation of the goings-on in prai-
rie-dog towns, brought renewed respectability to naturalistic studies of
complex social organizations and the efforts to relate their details to environmental factors, particularly the role of predators and competitors. I think
very few ethologists of my generation, at least within the United States,
were not impressed and influenced by King's work.

Some of Osborne's guests were famous long before they came. Konrad
Lorenz, for example, was a giant literally and figuratively, with a thick
white beard and the red cheeks and twinkling eyes of a Santa Claus. His
appearance certainly matched the legend, and his arresting platform style
allowed no doubt as to the profound and far-reaching implications of his
research. An absence of supporting data was easy to overlook, given the
forcefulness of his presentation and magnetism of his personality. We became friends on the spot despite my earlier misgivings and remained so in
spite of continuing disagreements.

I think Lorenz's intuitive approach is perhaps best illustrated by an
episode that took place about 20 or so years after this first encounter. Lorenz
was lecturing at local universities, and thus Martha and I were able to try to
repay the hospitality he had earlier extended to us at his Bavarian institute.
I had just before his visit read his book, *Man Meets Dog,* in which he proposed a diphyletic origin for domestic canids. Some breeds were derived
from wolves, Lorenz argued, others from jackals. The former were wary
but loyal, the latter fawning and fickle. There were other differences, too,
but I've forgotten them. We raise Norwegian elkhounds, and this breed
failed to fit either of Lorenz's categories, or so we thought. Having never
seen elkhounds before, he played with ours and seemed thoroughly at ease
with them. "Well, Konrad," I finally nerved myself to ask, "how do you
reconcile this dog with your scheme?"

"Why is this a problem? It's obviously a jackal type," he responded. I
almost fell off the sofa in surprise. Our elkhounds are (I thought) like large

**Konrad Lorenz in 1961**

gray chows, and chows are Lorenz's paradigmatic wolf type. Lorenz thought our dogs resembled the little barkless African basenjis.

"Konrad," I remonstrated, "the only thing elkhounds have in common with basenjis is a curly tail."

"Not so," he reported, "they have a similar gestalt."

"How 'similar'?" I demanded. "Body shape, limb proportions, head form, all differ."

"Peter," came the fatherly reply, "gestalts cannot be analyzed, they must be perceived."

We argued for quite a while about whether gestalts were subject to quantification and precise description. About the time I came to cite D'Arcy Thompson, Lorenz's patience came to an end. Banging his coffee cup onto the table (it separated from its handle), he ended the dialogue: "Peter, to see a gestalt you must have a clinical eye. You do not have the clinical eye!"

Another visitor to Osborne, this one from the medical center across town, excited less admiration than had Lorenz. He was Jose Delgado, an articulate, urbane, versatile, and skilled neurophysiologist, who was exploring the brain's circuitry and discovering ganglia that, on being stimulated, could excite or suppress specific behavior patterns—feeding, attack, or sleeping. There was nothing new about such findings per se, though controversy as to their significance for the understanding of brain function continues to this day. Delgado's contribution was the application of those findings to the control of human behavior. He had available to him patients in Yale's psychiatric ward in whose brains he could implant electrodes. His lecture ended with an appalling photo of a patient with two antenna-like structures protruding from his turbaned head. "Some day," Delgado cheerfully prophesied, "we will be able to control behavior."

It was my task to write summaries of Osborne seminars for publication in the august *Yale Journal of Biology and Medicine*. This seminar was too much for me, and I refused to submit it. I was much relieved when the editor agreed with my decision, though, in retrospect, it was the wrong one. Suppression is never a solution, and I should have known that.

Delgado went on to achieve popular fame when he was pictured in a suit of lights in a Madrid bullring. The bull made several charges, but each time it seemed Delgado's end was near it dropped to its knees, literally bowing at Delgado's feet. The bull carried an implanted, battery-operated receiver attached to electrodes pinned to a certain area in its brain. A small receiver in Delgado's hand did what other matadors had to achieve with physical skill and dexterity.

My last meeting with Delgado, who left Yale to return to Spain, was at a NATO conference on aggression held in the South of France. We both were on the program. Delgado in his talk described patients who were subject to violent outbursts of rage that led them to attack whoever was near. He discovered that these outbursts were preceded by a peculiar set of impulses that could be recorded from near the brain stem. He arranged a circuit such that brain electrodes fed impulses to a microtransmitter implanted under the scalp, which relayed the signal to a computer in his lab. The computer reacted to the appearance of the rage signal by activating another transmitter whose signal was received by an implanted receiver, which, in turn, stimulated electrodes implanted in an inhibitory region of the brain.

Delgado's patients could now, for a time, walk freely along the streets of Madrid without danger to others and without outward signs of the potent electronics hidden beneath their scalps. Delgado spoke eloquently and passionately about the prospects for behavioral control this technology posed. I was shocked that, in contrast to my colleagues at Yale, this audience (psychologists, neurobiologists, and ethologists) applauded enthusiastically. This time I spoke out in protest, arguing that for many of us, science is an intensely personal affair—not only a profession but also a vocation, in the ecclesiastical sense. Vocations, however, often generate an evangelical fervor, which may include an obsessive desire not merely to change the world, which is laudable, but also to control the world, which is not. When the scientist seeks control, the science ceases to be a personal affair, hence this comment: I was dismayed that at this conference, where our ignorance about aggression had been a continuing theme, we were still prepared to discuss means of control. I believe that anyone to whom the concept of the good in humans, or human dignity in the secular, is meaningful must voice a profound objection to experimentally motivated manipulations designed to irreversibly alter human behavior.

Delgado's quiet reply was: "It is spurious to argue that changes in the brain should not be allowed . . . besides, the patients or their kin gave us their informed consent."

Years later I was part of a group of neurobiologists assembled at a *parador* near Seville by the Frank Guggenheim Foundation. We had come together to review the very latest techniques in brain research directed toward understanding the organization and control of aggression. Seville had several years earlier been the site of the UN's conference on violence and aggression, so this venue had symbolic importance. It was also a lovely, relaxing locus, though that did nothing to diminish the fervor of the arguments, which returned to the theme Delgado had championed: the direct control of centers of aggression. My view, however, continued to be that the notion of "centers" was unsupported at worst and simplistic at best: we were still too ignorant.

Returning to Osborne, another of its notable guests was Margaret Mead. She was already a legendary force at that time, and I had been an avid reader of her numerous books and articles from the time of her first Samoan studies. She was acquainted with Hutchinson and evidently consulted with him on biological issues. One April morning as I slithered through the closing door of the elevator, I confronted my professor and his companion, a dowdy person he introduced as Mead. Portraits of her as a young woman which had formed my image of her, had revealed her as uncommonly beauiful, but I had no time to reflect on the changes time produces. "Young man," she ordered, "I want you to explain imprinting to me. Professor Hutchinson says you know about it." We agreed to meet after supper.

It had happened, however, that my wife had given birth to our first child only a few days earlier. I was loath to give up the only time of the day when we could be together. Martha saw no reason why we couldn't all three make the postprandial trip back to the lab. Thus, while Margaret and I compared notes on mammalian maternal attachments and their relation to the phenomenon that Lorenz had popularized in waterfowl, Martha unselfconsciously nursed our tiny daughter.

In the early fifties, breast-feeding was not a common practice among the New England middle class; it was certainly uncommon in a laboratory during a scientific discussion—even one on maternal attachment. We were unaware of the irony (or appropriateness) of our behavior, but it was evidently not lost on Mead, a long-time advocate of natural child-rearing practices. Years later she remarked that her continued interest in me was at least as much occasioned by her attraction to Martha as by what I had to say.

What we discussed, in part, was a view of imprinting that has had to be changed considerably as ever more experiments have been conducted.

The term "imprinting" is a translation from the German of *Prägung* and implies the impressing of a pattern upon a temporarily malleable substance (the analogy is with warm sealing wax), which, on hardening, retains the impressed pattern. It was employed by the German ornithologist O. Heinroth and more carefully defined by Konrad Lorenz to describe the behavior of young ducks and geese. Soon after hatching, these animals will follow a wide range of moving objects (normally their mother) and as a consequence of this first following (even for only a few minutes) come to subsequently prefer objects of the same kind: if the gosling first followed Lorenz, it would subsequently select Lorenz over other geese. Lorenz believed that the sexual preferences shown by imprinted animals were permanently fixed by their early experience. In short, imprinting was characterized by the early age and rapidity with which it occurred, the absence of conventional reinforcers (the following response did not have to be "rewarded" by being followed by feeding or warming), the generic nature of the attachment (if the object followed was a Greylag goose, any Greylag goose would subsequently be accepted, not a particular one), and the preference was irrevocable. In his book *Imprinting* (1973), Eckherd Hess details several decades of research that forced qualifications to these generalizations. But that came later. At the time, the imprinting process seemed ideally suited to species that were motorically precocial—i.e., could independently leave their parent, but still needed looking-after. It assured a rapid learning of the species' hallmarks, a process unnecessary to a robin or bluebird, which are physically compelled to stay put for many days after hatching.

Margaret Mead saw similarities between certain attachment processes Freud and his students had postulated to occur in human infants. I was more impressed by the ecological argument (that only precocial species, not humans, should manifest so risky a process), and that was the focus of much of our talk. That particular point is still unresolved, even while the distinctions between imprinting and other forms of learning have disappeared.

This was the first of many subsequent meetings with Mead, who became a great, though somewhat intimidating, favorite of my daughters. It also brought me and many of my students into contact with age-mates in anthropology, sociology, and medicine. Margaret fostered close relations with a covey of young researchers in these areas, and through a multitude of informal conferences and meetings that she organized, she arranged for us to meet and stay in contact. I think there is little doubt that much of the interaction that initially developed between ethology and the social sciences was promoted by Mead's mentorship of this heterogeneous collection of young investigators.

Don Griffin was perhaps less well known to the public at large, but to the biologists at Osborne he was already in the fifties a demigod. While still an undergraduate student, he had resolved a conundrum that has long excited attention: How do bats avoid obstacles even in complete darkness? That they do so was demonstrated over two centuries ago by an Italian investigator, Spallanzani, who had strung fine wires from the ceiling of a room. They were just far enough apart that to pass between them a bat had either to fold its wings or turn sideways. Bells attached to the wires signaled to an observer when the bats, moving in total darkness, had missed the mark. They rarely did. In 1938, Griffin together with his friend Robert Galambos, solved the mystery: bats avoid obstacles by hearing the echoes of ultrasonic pulses they emit while in flight. If the bats' muzzles were taped shut so that they could not emit these pulses, they would not voluntarily fly. If thrown into the air and thus compelled to fly, they collided with obstacles in their path. Occluding an ear had the same effect (Griffin 1958).

What set Griffin apart from many other physiologists who examine the relation between sensory capacities and behavior was the degree to which his conclusions were tested under naturalistic conditions. Thus, after carefully characterizing the behavior of bats flying in his technologically sophisticated studio, he developed portable recording gear and repeated his studies on free-flying bats. In this, as in many other ways, Griffin's work was important to the development of ethology as a discipline.

After Griffin's visit, I was emboldened to write him for advice on an unsuccessful project with which I had been toying for several years—ever since, in fact, I had been, as an undergraduate, introduced to Griffin's work. The project was an examination of fly catching by swallows, my assumption being that they, too, used ultrasonic echoes. Some years later, in fact, it was shown that some cave-dwelling birds (the S. American oilbirds, *Steatornis* sp) do echolocate, but evidently the swallows I was interested in do not. In all events, I was technically unsophisticated in acoustic engineering and instead of doing some homework wrote Griffin for help. I am embarrassed by this recollection, for when I receive letters of the genre, "Dear Professor: Please tell me about imprinting so I can write a term paper," my reply tends to the sarcastic ("Wouldn't you rather just have me write the paper?"). Griffin, on the other hand, responded with a multipage explanation of what I was doing wrong and how to do it right. He concluded with an invitation to visit his lab. While I was there, for I came apace, he glanced out his window and noticed a young woman playing with a puppy—Martha with our first Norwegian elkhound. Discovering her identity he called her in and then proceeded to repeat for her the hour-long lab tour he'd already provided me.

Judging from the many published testimonials that have appeared, Griffin's consideration for colleagues and students appears to have had no more bounds than his enthusiasm for and originality in research.

Griffin's interests included practical and social problems, and this was not altogether common among academics in the fifties. It was important to me, however, for I had been spending a fair bit of my time outside of the lab working with an interdenominational group ministry. Indeed, for the first two years of graduate school, I lived within the parish in a condemned house lacking plumbing or power. (Actually, the kindly prostitutes in the house across the alley from me provided a lamp and extension cord. However, about every two weeks or so their establishment would be raided and I'd be lightless for some days until my friends were bailed out again.) The others in the group were ordained ministers or students at Yale Divinity. As a Quaker and scientist, I was definitely odd man out, so to me fell the assignment of organizing collegiate weekend work camps for the repair of the most serious problems in our neglected neighborhood. This detracted from my studies, but it was an important activity for me, and I was relieved to have so successful a scientist as Griffin look favorably upon activities that some of my teachers deplored as "ascientific." My colleagues in the ministry provided a vital balance between the ordered, arcane world of Osborne and the unpredictable yet routinized life of most on this planet. This balance seems to be especially needed by ethology students, whose work inevitably forces them to consider, at least in passing, the foibles of human behavior and their courses. Perhaps that is why ethology students seem to have been overrepresented (relative to say, cellular-biology students) among civil rights demonstrators in the sixties, or the antiwar protesters of the seventies.

Reflecting upon those I've mentioned above reminds me of how unpredictable the long-term consequences of their ideas have proven to be. Some, such as Lorenz, forced their views repeatedly into center stage. His charm and eloquence allowed them to remain long after more sensible rivals were on hand. Others—Delgado, for instance—reflected an already popular view and were able to propagate their dogmas by trading on this popularity. MacArthur and Hutchinson, in contrast, produced work of such dazzling originality and careful documentation that their studies were repeatedly the impetus for other work, even though neither was particularly assertive, charismatic, or self-conscious about his fame. Finally, there were the quiet, less-visible workers—Collias, Liddell, Caspari—whose contributions were made known largely through the work of their associates and students, rather than through their own ebullience or the startling nature of their findings. Whose ideas will still be seriously discussed in future decades? For better or worse, it seems as if representatives of the first two groups are most likely to garner support and acclaim within their lifetimes.

# 3
## Cambridge Idyll

In the fifties, there were few labs to chose from if one wanted to do postdoctoral work in ethology. One of these was Bill Thorpe's Madingley Field Station for Animal Behaviour in Cambridge. Given Hutchinson's strong ties to that part of the United Kingdom, there was little doubt that this was where I should go. Peter Marler, now the premier specialist in birdsong, had just completed his studies there, and in the summer of 1957 had taken his wife Judith and his birds by freighter to California en route to a position at U.C. Berkeley. We met when his ship docked at Los Angeles and, over a picnic lunch in Griffith Park, discussed the advantages of working with Thorpe. Thorpe's focus was on song learning in passerines; my thesis dealt with how ducklings learn their parents' calls, so a certain symmetry in interests existed. When Marler offered to arrange for us to take over his previous apartment in Madingley, our decision was made.

Madingley, about three miles from Cambridge, looked like a restored "traditional" village. Its major landholder had been the owner of Madingley Hall, an ostentatious vestige of England's glory days and now a university hostel. The villagers lived in a double row of thatch-roofed cottages, buried in gardens of the largest, brightest flowerbeds we'd ever seen. There was, naturally, a pub at the entrance to the one street, and the field station, a collection of ramshackle huts scattered about several acres of woods and fields, lay conveniently behind it. Opposite the pub's end of the station lay the grange, formerly the gamekeeper's abode and home to our landlord (a Cambridge biochemist), his wife, and their two babies. A more hospitable home could scarcely be desired.

MacArthur, my close friend from Yale, elected to pursue his studies with David Lack, at the Edward Grey Institute of Oxford. Thus, MacArthurs and we, our two yearlings in tow, sailed together in the newly christened

**Madingley Village**

**The field station at Madingley in the late 1950s**

*Stattendam,* a Dutch liner that offered respite from the stress of thesis and thesis exams, and provided frequent and highly caloric meals.

This was the first such voyage for any of us, and, perhaps unfortunately, it was so idyllic, the crew so helpful and polite, the ship so comfortable and polished, that an all-too-high standard was set. (We realized this one year later, when we returned to the States on a French liner, grimy and unwashed, peopled by a surly, hard-to-locate crew. The dining-room stewards wanted children excluded, so when we came to dine, we were punished by being made to wait until the end to be served. Many items—including that childhood staple, milk—were then no longer available. The stewards did not allow time enough for their consumption, anyway.)

**Robert and Betsy MacArthur, PK, and our firstborns on the river Cam in 1957**

Madingley Field Station was William Thorpe's creation. He was master of Jesus College and author of what was, up to that time and for a decade after, the most encyclopedic and influential book on animal behavior (*Instinct and Learning in Animals,* 1956). His earlier work had been in entomology, where he had worked on the problem of host preferences and specificity. This included the discovery of larval conditioning (or olfactory imprinting), a process in which the food devoured by the larvae determines the plant on which they will preferentially deposit their eggs when they reach adulthood. The full ecological and evolutionary significance of this phenomenon was not apparent until years later, but it alerted Thorpe to the potential importance of brief experiences early in life. The learning of birdsong by inarticulate nestlings seemed an analogous phenomenon, so Thorpe's transition from insects to bird work was not surprising.

Thorpe was an enigmatic man to many of his associates and students. He was painfully shy, and on first entry into his presence, greetings and pleasantries exchanged, he was wont to sit silently, waiting for his guest to take the conversational initiative. Younger students, already in awe of this famous figure, often found this difficult. Thorpe, however, like I, was a Quaker, and so far from finding the silence intimidating, he appreciated being allowed time to allow the appropriate thought to develop.* Our conversations were frequently punctuated by silence, and, as we came to know each other, I think we both found the pauses congenial. They were a cause of embarrassment to many of my peers, however, who found Thorpe stiff and formal—which was not the case. It's unfortunate that his record as a postdoctoral student at the University of California (Riverside) was not better known, for it would have softened students' view of him.

As a postdoc, Thorpe fell in love with the austere beauty of the nearby Mojave Desert, and, along with friends bemoaned its desecration by the proliferation of billboards along the roads that dissected it. He and his friends organized to take direct action, in the manner of Abbey's Monkey Wrench Gang, but antedating Abbey by half a century. One particular billboard, which advertised a hotel in downtown Los Angeles, defied the saws and torches of the ecopirates: it was assembled of steel beams. Thorpe became an expert in the use of socket wrenches. To avoid detection, however, he had to proceed carefully, for the disassembly took considerable time. His procedure was to loosen the bolts so the sign would not topple immediately, but only after the wind had played with it for a bit.

A windy night was nearly his undoing. Police had been on the lookout for the billboard vandals when Thorpe, his last nut loosened, joined the pickup crew for the ride home. The car stalled and would not start. The police arrived in due course, but as there was no evidence of a vandalized sign nearby, gave assistance. The engine finally started, and the police departed; behind them were Thorpe and his friend, buoyed by a great gust of wind that brought the sign crashing to the ground. The hotel it advertised was the Los Angeles Biltmore, Thorpe told me.

---

* —reminding me of a sign posted in a Friends school: "I am a Quaker. In case of emergency, be silent."

Thorpe, along with his many academic and scientific responsibilities, was also an active and influential member of the Cambridge Friends Meetings. His activities in the peace movement of the sixties and seventies gave heart to the many of us whose junior status was often seized upon as reason enough to ignore our concerns.

We were awakened on our first morning in the grange by the incongruous wail of Abrams' four-year-old: "Mommy, come wipe my bottom," an oft-repeated call that soon became our family's wake-up signal. Breakfasting by our kitchen window, we were entertained by tits (*Parus* species) of different colors and sizes pecking through the tops of our milk bottles to slurp up the cream. Robert Hinde and James Fisher had just recently published an account of this behavior and had tracked its spread through Europe. Evidently, the trait had developed spontaneously in several localities and was then centrifugally spread through observational learning: "Each one teach one".

Martha subsequently noticed that the birds exhausted the cream from bottles topped by gold foil before they attacked those with silver. Red-topped bottles were often ignored—those contained skim milk, while gold and silver signaled cream and whole milk, respectively. She conducted a study that confirmed the birds' knowledge of this scheme of things. By switching tops she could switch color allegiances. It seemed as if one experience sufficed, but learning was complicated if several birds were around because they apparently learned empathically, by watching one another, as well as directly.

During the course of our breakfast-with-tits, the first of many, a knock on the door brought us to face-to-face with Robert Hinde. It was a lovely coincidence, and a special delight to meet Hinde, who soon came to symbolize for us the friendly, gracious aspect of Cambridge life, to say nothing of its intellectual breadth and intensity. Hinde was still conducting his well-known studies with canaries on the interactions of exogenous and endogenous factors in the control of mating and incubation, studies that in many respects paralleled those of Danny Lehrman in the United States and that did much to confirm the ideas developed earlier by Frank Beach. Hinde's great erudition and meticulous scholarship influenced all who read his work, but his difficult-to-read style has regrettably cost him many readers.

Several of the students at Madingley worked with Hinde rather than Thorpe. Foremost in my recollections is Thelma Rowell, then studying reproductive and parental behavior in hamsters, and since then renowned for her work on social behavior and development in monkeys. (Hinde, too, later moved to studies of primates and in his elder years, human primates.) Her work on monkey societies remains a classic in its field. Hugh Fraser

Rowell, since turned to neurophysiology, was still a behavior student with Hinde at the time. I believe it was he who taught me that the time required for finches to overcome their fear of an observer and accept feed in his presence varied directly with temperature: the colder the day, the more quickly (or closely); this became a useful stratagem for some of my later studies. Robert Klopman was a Thorpe student, along with Janet Kear. Both were involved with studies of waterfowl—useful to me, as I needed help finding duck eggs for my imprinting studies—and for many years they contributed to studies of goose behavior and avian orientation, respectively.

Malcolm Gordon, a comparative physiologist, was then working out the different mechanisms for maintaining isotonic equilibrium used by trout inhabiting the sea and freshwater streams. This continued to be a major theme of his for many years, though most students probably know of him through his encyclopedic but readable physiology text. At Yale he was regarded as unduly aggressive and avoided by many of his peers, but the Cambridge ambiance had softened some of the edges. Later, his feisty inclinations were of enormous value when he became a leading figure in the Southern California American Civil Liberties Union.

Malcolm's lab mate at Cambridge's Downing St. labs was a muscle physiologist and avid ornithologist named Colin Pennycuik, as quiet as Malcolm was energetic. The three of us occasionally rattled around in Colin's old car to view birds, particularly at the rich sewage farms outside the town. Colin went on to Nairobi; we met there years later and I was treated by him to an airborne tour of the East African plains. Colin was addicted to gliders and small, light aircraft and apparently thought nothing of cruising alongside a herd of galloping giraffes, eyeball to eyeball, at giraffe-height altitude. Later, Colin combined his avian and aviation interests with his physiological skills and became renowned as an authority on the dynamics of flight and soaring in birds.

There were not so many Americans at Cambridge in those years that we were likely to randomly encounter one another. Our English colleagues, however, seemed to think we required mutual consolation, for they never failed to alert us to the presence of a countryman. Thus we came to meet Steve and Ruth Wainwright, newly arrived from North Carolina's Duke University, an institution whose name was then unknown to me. Steve was completing a second B.A. at Cambridge, from which he would go to U.C. and then, finally, end as a titled professor at Duke. On that first, informal meeting, when we exchanged views on the future of our subject and of our own dreams, Steve remarked casually that it was his hope to be a modern-day D'Arcy Thompson, the famous morphologist. It was a long, slow start, with a lengthy excursion into the plastic arts and sculpture (from which some thought he might never return), but with the founding of his Institute for Biodesign at Duke and the publication of his books, Wainwright can indeed be regarded as the modern Thompson. His is the most singular instance I know of a young scientist recognizing a particular star and following it steadfastly and creatively for over four decades.

The roads between Cambridge and Oxford were not direct, and the trains were slow and costly. I suppose that is why there was so little direct communication between the behavior and ecology students of the two schools. It seemed a pity, but since I had personal reasons for making the tedious trip, I did not let past practices stand in my way. The MacArthurs were resident at Oxford, and Martha and Betsy were as eager to get together as were Robert and I.

Robert's workspace was in the Edward Gray Institute, headed by David Lack, which was already a world center for ornithological field studies. Lack had been a secondary school teacher and amateur ornithologist until he leapt to fame (and the call to Oxford) with the publication of his *Life of the Robin,* one of the half-dozen or so books I still insist my students read (the two others of equal priority being Fraser Darling's *A Herd of Red Deer,* and D'Arcy Thompson's *On Growth and Form*). I had been much impressed by the personal concern Lack displayed towards his students, as evidenced by his reception of the MacArthurs. On our landing, complicated by our frantic search for some N. American Warblers that had made the trip with us, a special messenger brought Robert a welcoming letter from Lack. Among other things, it offered this apologetic advice: "You may find us unfriendly and distant, but that is not the case. It is just that this is a very crowded island and only by erecting walls and fences can we maintain equanimity." I often remembered that advice when, at Cambridge, I was admonished for continually greeting people when I passed them in the halls. "If you've nothing of import to say," I was told, "don't interrupt the silence."

This friendly concern was immediately evident, when Lack dropped what he was doing to take his unannounced visitor on a tour of Oxford, highlighted by the adventuresome ascent to the Tower of Swifts. His work on the Galapagos finches and, more importantly, on the mechanisms of population regulation incited as much work in population ecology and the related areas of behavior as did that of any other single man. And, as with Lashley, it was work produced by a modest man whose ego never seemed to be a driving factor. Had he remained a schoolmaster he likely would have lacked the means to be as productive as he became at Oxford, but I do not doubt he would have worked as hard and as creatively even had he not been publicly recognized.

The Bureau of Animal Populations, next door to Lack's lab, was a significant center for MacArthur, but I was too unsophisticated mathematically to be able to follow many of the discussions there. I did come to know Monte Lloyd, who spent a great deal of time with MacArthur, and whose studies of the causes of the cyclicity in periodic cicadas have achieved some fame. Our paths sometimes crossed in Panama and North Carolina, where some of his cicada populations reside. Unfortunately, with cycles of emergence of thirteen or seventeen years, he doesn't get back to N.C. very often, but when he does the memories of Oxford of the fifties return in full flood (see Masterman 1952 for a hilarious but accurate account of life at Oxbridge.)


### POLITICS: MARXISTS, NATIONAL SOCIALISTS, AND OTHERS

During this period (the 1950s), it was not rare to encounter scientists whose directions, methods, and conclusions were influenced if not dictated by political considerations. The lessons from Germany in the forties had not been persuasive—as the Russian example later showed us again. Indeed, even today, in the 1990s, when many scientists would claim their work to be politically neutral, merely the methods of grant distribution, to say nothing of congressional and agency constraints, assure that party lines are recognized and followed. However, in the United States of the fifties, the attacks on scientists (and others) by Senator Joe McCarthy, and later then-representative Richard Nixon, helped to generate the counterclaim that one's political beliefs were irrelevant to one's scientific pursuits. It was a view I was happy and naive enough to accept. It was at a seminar for students of behavior and neurobiology that this thoughtless drifting struck a rock. The Cambridge neurobiologist J. S. Kennedy was defending his model on the organization of the motor-control mechanisms for insects. In repudiating an alternative model he matter-of-factly declared, "Of course, as a Marxist, I couldn't possibly accept that view." Aside from the shock of hearing someone publicly admit to being a Marxist, I had never considered that a creative and credible scientist would be consciously driven by political ideologies.

Kennedy's statement lent fire to countless later discussions: Could we, should we, attempt to separate our science from other aspects of our personal lives? Thorpe seemed to me to provide a model of a successful duality: a politically neutral science complemented by a strong and faithful commitment to a particular social and political platform. This was the model that appealed to me (and that, for better or for worse, I tried to follow for most of my life). Over the years, however, other models became familiar. A notable one was that provided by Warder Clyde Allee, whose methodology was "neutral," but whose choice of problems to investigate was not. A Quaker, Allee focused his attention on the adaptive value and evolution of cooperation and sociality.

The observation that science is influenced by politics has been noted since the days of Galileo, but the details as to how, by whom, and to what ends differ so much from case to case that the theme remains interesting. During the cold war it was, usually, physics and chemistry, occasionally mathematics, whose directions were thought to be influenced by political pressures (Snow 1961). Biology came into prominence with the Vietnam War, and interest in an array of biological weapons, from defoliants to nerve gases, likewise influenced a great deal of research. If one's memory goes back to earlier times, one also recalls the relations that developed between psychology and the politics of immigration and education, which had a lasting impact on developments in the study of intelligence (Gould 1981). Nor have the politics of religion been irrelevant (Durant 1985).

Ethology, as a coherent discipline, received an ex post facto baptism with the award, in 1973, of the Nobel Prize to Konrad Lorenz, Niko Tinbergen, and Karl von Frisch. The award followed four decades of research: by von Frisch on the complex system of communication among honeybees, by Tinbergen on the hierarchical structure of instinctive behavior, and by Lorenz on the mechanisms that underlie instinctive behavior. It was Lorenz's concept of the releasor and its associated innate releasing mechanism that for nearly forty years dominated the ethological landscape and influenced the directions and content of ethological research and theory (and see Kalikow 1983).

Lorenz, in his Nobel address (1974), claimed that his primary motive was systematics: he wished to use behavior as an anatomist did bones in order to both reconstruct phylogenies and to infer the functional significance of stereotyped movements (cf. Podos 1994). His observations focused especially on waterfowl, and particularly their courtship rituals. In the course of these observations, he noted that displays sometimes occurred when the usual eliciting stimulus was absent, or was present in a distorted form. He concluded that the threshold for the elicitation of some stereotyped behavior patterns or displays must fluctuate. The model he devised to account for this is now enshrined as the hydraulic or toilet-bowl model. I have not tried to estimate what percentage of the pages of *Zeitschrift für Tierpsychologie,* the publications of the venerable British Association for Animal Behavior, and *Behaviour,* the three main outlets for early ethologists, were devoted to studies of these purported mechanisms. It was surely significant. Nor could we describe the content of ethology without reference to IRMs, RMs, SAPs, FAPs, and other acronyms that refer to attributes of the model.

The resistance to the analysis of behavior in terms of the hydraulic model was galvanized by a critique by Dan Lehrman, published in 1953 in the *Quarterly Review of Biology*. In his arguments, Lehrman, along with criticisms of Lorenz's methodology, hinted at other than empirical influences at play in the construction of the model. Lehrman and his associates posited that Lorenz had Nazi sympathies, which were revealed in two articles not listed in Lorenz's bibliographies for many years. These were "Die angeborenen Formen moeglicher Erfahrungen" (1943) and "Durch Domestikation Verursachte Stoerungen" (1940). In them, Lorenz argues that it is important to prevent interbreeding of persons of different so-called races (it must be noted that the German concept of race bore little relation to what most anthropologists, and certainly biologists, understand by the term). Basically, he claimed that since displays of waterfowl are species-specific, hybridization destroys the integrity of the releasor mechanism and leads to the destruction of the species. By analogy, humans are believed to possess releasors for ethical and aesthetic values that also are lost with "hybridization." The lack of vigorous selection under conditions of domestication also allows the proliferation of the "Minderwertig" (inferior), who ought to be "ruthlessly extirpated" (1940, 66).

After the war Lorenz emphatically denied he had had Nazi sympathies, and explained the offending articles as a naive effort to obtain and then retain an academic post in difficult times. By 1943, when the second article appeared, he had become professor at Königsberg. Wieck (1990) has pointed out that at that time and place, no one, certainly not Lorenz, could have been unaware of the policy of "euthanasia" of the physically and mentally infirm, which the Nazis had initiated even before the establishment of their death camps, and which his 1940 paper urged. Yet, Tinbergen, imprisoned by the Nazis as a hostage against the Dutch Resistance, and von Frisch, himself once the target of the Nazis (he was spared, it is claimed, by virtue of the economic importance of his research on a virus that infected and destroyed bees) were, after the war, reconciled with Lorenz. Discussion of the matter was dropped.

While preparing a biography of Karl Buehler, the Austrian psychologist, the linguist Thomas Sebeok displayed to me a letter he had found. In the spring of 1938, Buehler had been arrested by the Gestapo and held for several weeks before being released without charges. The letter in question was the copy of a craven note he had written to the authorities, thanking them for the opportunity they had provided for him to reflect upon and reform his ways. Heretofore his work had been apolitical, Buehler wrote, but in future it would advance the goals of the Reich. In his defense, he added, he had, though himself apolitical, shielded from prosecution many of his coworkers who had been early members of the party in Vienna. Konrad Lorenz was among those listed. Shortly after, Buehler fled to the United States (Sebeok 1981).

This letter, which seemed to contradict Lorenz's claims, prompted me to examine some of Lorenz's prewar correspondence, especially that between him and his mentor, Oskar Heinroth, the world-renowned ornithologist and director of the Berlin Zoo. Heinroth was evidently no friend of the Nazis. His letters betray no sympathy for the Third Reich and I never saw a letter ended by him with the then-customary salutation, "Heil Hitler."

The letters deal mostly with the behavior of the ducks and geese the two friends, Heinroth and Lorenz, were regularly exchanging and studying. Interspersed (and conspicuously absent from the published collection of these letters [Koenig 1988]) are political asides: references to Lorenz's impatience for a war with England so that "that arrogant race can be taught a lesson" (18 December 1939, Nachlass 137, Ordner 27; see Koenig 1988 for the full catalog reference); anti-Semitic jibes, as when Lorenz describes the shoveler duck with its "ugly Jewish nose" (21 January 1939). More significant, however, was the correspondence that preceded the publication of the two articles to which Lehrman had in 1953 first called attention. Lorenz, as various of the letters show, clearly knew that, different releasors or not, viable duck hybrids between species (and even genera) could be formed. One must recall that the hydraulic model, which formed the basis of Lorenz's theory of instinct, depended on the specificity of the innate releasing mechanism, a lock that only a specific key, or releasor, could open. Lorenz's speculations on different human physiognomies and standards for beauty and ugliness—his association of the proud and the beautiful with Aryan ideals and the inferior with urban Jews and Gypsies and other decadent products of domestication—are repeatedly voiced in manuscripts or letters to Heinroth in 1938, 1939, and 1940. Occasionally, the arguments on *racial* standards of beauty are transmuted into discussions of preferences of the species.

A word about biology in the Third Reich is in order here. It was accorded a higher priority in the schools than all other sciences. Under a fervent Nazi pedagogue, Ferdinand Rossner, the new NS biology became a central theme in all German schools and was accorded time taken from other fields, especially foreign languages and mathematics. New texts were quickly introduced to enhance the curriculum with a heavy emphasis on *Rassenkunde* (the science of the races). Lorenz was evidently among the best known of the scientists who contributed to these developments, a fact only recently revealed to the English-speaking public (Baeumer-Schlein-kofer 1992a, 6).

Contrary to Lorenz's previous assertions, he did not go directly to the Eastern front but first served (1942) in Posen as a psychologist with an SS unit assigned to perform tests that would allow distinctions to be made between Poles and Polish-German "hybrids." He was a member of the Rassenpolitisches Amt, the "race" division of the SS.

In sum, the ideology of the NS state required biological substantiation, and this Lorenz provided in greater measure than most other biologists of note. His motivational model and its application, I suggest, may have been derived as much from this ideology as it was from his studies of animals. Indeed, many of the observations he submits to Heinroth's criticisms even contradict the predictions of the model. However, the popularity of his books, and the charisma of the man himself, diverted attention from his past, and he became widely loved and honored, and his work, until very recently, generally accepted (but see Zippelius 1992, who claims that the results of many of his studies were "fudged" to fit Lorenz's preconceptions).

Perhaps, at this point, it is best to let the words of Lorenz speak for him. From a 1940 paper, never before published in English, Lorenz writes:

[S]hould there be factors that favor mutations, then the most important task of those who protect our race would be, first, to recognize, then to eliminate them...But if it should turn out that under the conditions of domestication no accumulation of mutations is taking place but that, instead, through the mere omission of natural selection, the number of mutants present increases and an imbalance of ethnic groups develops, racial hygiene must nevertheless aim toward an even stronger elimination of racial inferiors than is presently the case; in fact, under these conditions, racial hygiene literally will have to substitute for all selective factors that ordinarily take care of selection. In addition, even after successful elimination of deficient mutants, an ever watchful selection process must be maintained, ready to deal with a renewed appearance of homologous mutations.

If a given [inferior] ethnic group should suffer even the slightest deficit of social inhibitions with which they may be afflicted and which would prove disadvantageous to them were they to live in an agricultural or fishing village, [such weakening] may enable them when living under the conditions of a metropolitan city to take advantage of ethnically valuable people, thereby becoming dangerous parasites to the entire nation. Everybody is capable without limit of visualizing many such cases that have actually occurred, where socially valuable traits of those with high [racial?] qualities were by such means "rewarded" with a negative selection. Wherever competition for space among conspecific groups is the only operative way of selection, the phenomenon referred to above will have the result of enabling socially inferior human beings precisely through their inferiority to invade the healthy body of the nation, and eventually to destroy it. Using the farthest reaching biological analogy, this is the same process in which cells of malicious tumors operate within the cellular body of higher organisms. . . .

From the far-reaching biological analogy of the relationship between body and cancerous growth, on the one hand, and, on the other, a nation and those of its members that, on account of deficits, have become asocial, important parallels emerge as to the necessary measures to take. Just as with cancer—apart from some insignificant partial results of treatment by radiotherapy—suffering mankind cannot be given any other advice than to recognize the evil as soon as possible and then to eradicate it, so racial hygienic defense against elements afflicted with deficits is likewise restricted to employ the same rather primitive measures. . . .

Every attempt to regenerate elements that have dropped out of their integral wholeness is therefore hopeless. Fortunately, though, their elimination is easier for a physician trying to treat national ills, and less dangerous for the superindividual organism, than are the operations by a surgeon. The great technical difficulty is to recognize them. But in this connection we can be greatly assisted by the cultivation of our own innate patterns, in other words, by our instinctive reaction to deficit symptoms. A "good man in his dark yearnings" (Goethe's *Faust,* Part II) is very well capable of recognizing whether another person is a scoundrel, or not. . . .

And yet, some human corporate body will have to assume this task if mankind, lacking the essential selective factors, is not to perish on account of its symptoms of decay conditioned by domestication. With the racial idea as the basis of our form of state, much along these lines has already been accomplished. The Nordic movement has always been directed instinctively against "turning human beings into domesticated pets"; all its ideals are of such a nature that they would be destroyed by the biological consequences of civilization and domestication presented here. It is fighting for a course of human development directly opposed to that in which the inhabitants of big civilized urban centers presently participate. Nobody with a sense of biology can have any doubt which of these two directions is the path of the real evolution, the way leading upward!

Therefore, the most effective measure to cultivate race, at least for the moment, is certainly the support of the natural defenses wherever possible; we must—and may—rely here on the sound instincts of the best among us and can entrust to them the selective process on which the success or doom of our people will depend. Should this selective process fail, should the elimination of those elements afflicted with deficit symptoms miscarry, then these [elements] will penetrate the body of the nation just as cells of a malignant tumor will penetrate a healthy body and eventually will destroy it alongside themselves. (Lorenz 1940; translated by W. T. Angress)

Michael Wieck, a literary critic (and German Jew who survived the war) places Lorenz's two tracts, "Disturbances Caused by Domestication" and "The Forms of Possible Experience" into their contemporary context:

What a verbose edifice of thoughts, the shaky foundation pillars of which are concepts, nowhere clearly analyzed, like persons "of perfect creation" [vollwertig], "ethically inferior," or "afflicted with functional deficits." Concepts such as these enabled unrestrained murderers living under a brutal dictatorship to justify their crimes. Furthermore, Lorenz gives the impression that it is possible to speak about the racial homogeneity of any European nation, and he breaks down people into those who are beautiful and good, or ugly and base. Over and over again he emphasizes the necessity of extermination [Ausmerzen], and in order to lend this appeal still more weight calls forth the terrifying specter of a body infested with cancerous cells. . . . For those who want to know how to recognize the cancerous cells not of foreign racial origin, i.e., those "afflicted with functional deficits," he has on hand an assertion rather peculiar for a scientist: "A good man in his dark yearnings is quite capable of recognizing whether another person is a scoundrel or not." The "selection," though, and the "extermination" process he leaves to the "best." Who that was during the Third Reich he postulates as being obviously well known. (Wieck 1990, 97; translated by W. T. Angress)

He continues,

Lorenz must have known in 1940 to what he lent his hand with his instigations. He unequivocally commended the racial policy practiced up to then and demanded an even more radical extermination of those deemed ethically inferior. Furthermore, Lorenz knew then full well that the racial policy of which he approved was directed against Jews, Gypsies, Negroes, Slavs, Enemies of the People (i.e., those in opposition to the regime), and handicapped and asocial people.

How many have taken the effort to investigate how it became possible that thousands of so-called normal individuals allowed themselves to be harnessed to a machine of mass murder? For all we know they came across articles such as this [Lorenz's]. Hearing arguments buttressed by references to cancerous growths, it was possible to turn off one's conscience and to believe the unbelievable. . . . Yet for me the unanswered question remains why those whose exhortations in fact led to these criminal actions were subsequently never called to account for what they did, but received on the contrary the highest honors, while those whom they seduced were executed or punished. Is it really so much less evil to want, to demand, and to justify that people be "exterminated" than to carry it out? These thoughts occurred to me long before the city of Vienna decided to make the Nobel Prize-winner Lorenz an honorary citizen. (Wieck 1990, 97–98; translated by W. T. Angress)

Lorenz's final decade closed on an ironic note. On his retirement he was summarily moved out of his Seewiesen home and lab. The Max Planck Gesellschaft had assigned it to his former student, Wolfgang Wickler, no disciple of southern courtesies, and Lorenz returned to his homeland, Austria. A controversy had developed there concerning the construction of a nuclear reactor. Konrad took the side of the conservationist opponents, a position fully consistent with his naturalist inclinations; as a consequence he became the darling of the local Greens. And the concept of the releasor, and its associated baggage, has proven to be a heuristic of inestimable utility.

# 4

## Duke in the Troubled Sixties

WHITES ONLY

Old-timers will remember Cambridge in 1958 as wet, windy, and generally sunless. On a short excursion to Norway that June, we were almost blinded by the sun when our ship passed the three-mile limit off the British coast and emerged from the fog within which we'd been living. Unfortunately, on our return about a fortnight later, we were as suddenly plunged into darkness by that same obstinately stable cloud.

At that time, I was watching birds from blinds, examining the process of observational learning in conditions as natural as I could contrive. I kept various species of passerines in large flight cages that could be divided in various ways. Thus, I could train individual birds to avoid seeds placed on certain backgrounds, but to accept them when placed on others. Bird "observers" could be allowed to watch the entire training process or else only the successful performances. It appeared as if some species only registered the successful responses of the "actor" bird, while others mimicked both correct and incorrect responses, seemingly ignoring the reaction of the actor (when a "wrong" choice of background was made, the bird received a seed whose kernel had carefully been replaced by aspirin powder, a nontoxic but, for birds, decidedly repellent substance). The difference in the manner in which observational learning took place seemed to be related to the social organization of the species, a theme I continued to explore for some years.

In 1958, however, it was cold work, and my frozen toes and fingers had frequently to be removed to the vicinity of a small electric heater in a nearby hut so they would be flexible enough to record data. Inevitably chilblains developed, and I alternately suffered demonic pains or distracting itches. When, unexpectedly, a phone call from Karl Wilbur at Duke

64

University came, all I could think to say was, what's the temperature there? It was in the 80s (Fahrenheit). I said I'd come.

I knew nothing about Duke, not even its location. Hutchinson had earlier written me about job openings at Amherst and Duke, and I had responded saying that I'd dearly like to be at Amherst. My heart was still in New England, and my frequent earlier visits to Amherst had left me impressed by the quality of its faculty and students. Much later I learned that Hutch had written a similar letter to Lincoln and Jane Brower, my Yale classmates, who were in England, too. Their response was complementary to mine: they preferred Duke, because its southern locale was more advantageous for their research on viceroy and monarch butterflies. Whether by accident or design, Hutch managed matters so that the Browers received an offer from Amherst and I from Duke. We both badly wanted to trade, but, in the end, I think were both satisfied.

Our flight that August from the United Kingdom to Durham was hardly of global proportions, but it was a change of worlds. The shock of arriving, in 1958, from cosmopolitan Cambridge to provincial Durham cannot be likened to a jump into an icy lake. Durham was hot and humid that summer, Cambridge cold and wet. Still, it was a shock. Durham's airport consisted of a single grass-lined runway, along which the baggage was dumped. The modest building (with a large sign proclaiming it an international airport) did, however, boast four separate restrooms, respectively labeled "Ladies," "Gentlemen," "Colored Women," and "Colored Men." The full significance of this dawned only some days later when we were rudely taught that launderette signs identifying certain machines as for "colored" referred to the user and not the clothes.

Our rural neighbors greeted us cordially the day after our arrival bearing gifts of food and helpful advice—but also expressing great relief at their discovery that we were not Catholic or Mormon. Being Friends (Quakers) wasn't quite as good as being Baptists, we were tactfully informed, but we would still be allowed to join the local volunteer fire department.

The segregationist ethic and religious intolerance, I soon discovered, was less gently defended at Duke. Within my first week, a senior colleague in the zoology department told me bluntly that my views on Southern practices were unwelcome; if I was to be another Northern troublemaker, I should leave then and there. Her attitude, fortunately, did not reflect the views of most others in the department, but was, I'm sorry to say, widespread elsewhere in the university. It was to be many years before I could look foreign colleagues in the eye and admit to being from Duke without an inward cringe.

A few years after arriving in Durham I received a phone call from a

Duke undergraduate I had come to know, Joe (Buddy) Tieger. Buddy had been involved in organizing those few Duke students who were prepared to participate in the sit-ins of Durham's segregated eateries.

The sit-in movement had begun in neighboring Greensboro in 1960, led by the students from the North Carolina A. and T. College, a Black institution. The students at a Black college in Durham (now N. C. Central University) soon followed suit in Durham, and a few Duke students participated. After being arrested, they were released singly, often late at night, and forced to run a gauntlet of segregationists gathered outside the courthouse cum jail. They were often singled out for special treatment, for town-gown hostilities at that time were considerable, and based on more than differences in racial politics. Buddy arranged for sympathetic faculty to meet the releasees and drive them to the relative safety of the campus (where they were then subject to disciplinary action for having been arrested). After a few weeks of escalating violence, a new and far-sighted mayor enlisted the support of community leaders and secured a moratorium on demonstrations. During this period a biracial commission was to develop a plan to end segregation of public dining facilities, and it actually succeeded. Thus, the focus shifted to Chapel Hill, home of the nation's first land-grant university, which had a much larger and (at that time) more politically active student body.

Buddy Tieger's organizational talents and dedication were soon called upon. The Chapel Hill student leaders wanted to impress upon the community that their views were shared by senior faculty. Buddy's task was to recruit faculty supporters who were willing to take actions that might lead to arrest and trial (to say nothing of mob violence). He recruited five from Duke, myself included. On the fateful day, we were joined by a pair from the University of North Carolina and a Black student, the faculty member from Central having been forced by his school's administration to beg off.

Our plan was to present ourselves at the lunch counter of a local motel/restaurant that was favored by Carolina students for postseminar snacking. We intended to argue that student and faculty patronage would not be adversely affected by the presence of mixed parties. If still denied service, we would decide at that time whether to go or stay, depending on our assessment of the situation.

We hadn't done our homework. None of us realized that the proprietor and his staff were stalwart supporters of the KKK. Neither had we known that a principal reason for the popularity of his facilities with Carolinians was that he rented his motel rooms by the hour. In the days before coed dorms this was no trivial matter.

Our arrival was anticipated. Most of our group never entered the restaurant, but were fallen upon and beaten to the ground as we emerged from our cars. Those who had gotten through the doors were dealt with especially harshly, but all of us were bruised or bloody before the police, who passively watched most of the proceedings, finally interfered and arrested us. Cold and wet—it was a freezing January night, and we had been hosed down as well as beaten—we were handed over to the jailer at Hillsborough's County facility (the Chapel Hill jail was already full), who opened the windows, turned off the heat, and left us for the night.

Sometime the next day we were released on bond and freed to contemplate charges of trespass, assault, and aggravated assault on a female, charges sufficiently serious that the magistrate before whom the case was brought bound us over to the Superior Court.

Our problems soon were compounded when the hard-line attitude of Chapel Hill's mayor and council led to an escalation of demonstrations, and considerable violence. A massive street blockade that snarled traffic after a basketball game heated the tempers of even those for whom integration/segregation was not an emotional issue. Defendants in sit-in cases, especially those who had been singled out by the press, had to keep a watch over their shoulders, and this was particularly true for Duke professors. Not surprisingly, no local attorneys would defend us, those from the NAACP excepted. We wanted counsel independent of the civil rights organizations, which were committed to defend individuals who were challenging the law by sitting-in. When we approached Watts Grill, it was not with the intention of breaking the law, but of seeking dialogue and conciliation. We were surprised and shocked when we discovered that we had a tiger by the tail, as it was put by Paul Hardin, then our advisor at Duke's Law School and later the chancellor of UNC.

The young lawyer whom Hardin finally found for us, Wade Penny (who, years later, became a leading member of the state house),was prone to classical quotations—no asset in front of Orange County juries—but was very much a local boy, very bright and dedicated to justice.

In the end, the trials of all the other defendants except for the Duke Five were removed to federal court. I recall Judge Mallard, an ironfisted, undersized man (on one occasion he asked an attorney to come closer, without irony remarking that "the Court was small of stature and could not see you from that distance"), admonishing the NAACP's attorneys Malone and McKissick that he had never treated them unfairly; why were they petitioning for removal to federal jurisdiction? His tirade over, Judge Mallard, other officers of the court, and the jury panels adjourned for lunch at the Colonial Inn. The Black lawyers and defendants (and a few others) lined up at the window of the town's hot-dog stand, the only "dining" establishment open to them.

The trials fairly quickly ended in convictions, except for mine. My refusal to be sworn, hand on the Bible, first led to an additional contempt of court charge, quickly withdrawn when Wade Penny presented the court with a copy of an old N.C. statute that specifically exempted members of the Society of Friends from the taking of oaths (North Carolina had been heavily settled by Quakers in the eighteenth century, and some of their influence persisted). This may have influenced the jury. In addition, the luck of the draw (and an oversight by the prosecutor) placed the daughter-in-law of a prominent Episcopalian priest on my jury. Her husband's father had himself participated in sit-ins.

After the presentation of witnesses, Wade appealed to the jury's sense of history. Hillsborough, after all, was where the Regulators first defied British authority before the Revolutionary War began. The battle against segregation was to be read in that light. Solicitor Cooper countered that the large oak outside the courthouse was where the Regulators were hung for their deeds. My jury hung, too, and I was rescheduled for trial at the next session.

In the interim, the U.S. Supreme Court ruled that the Public Accommodations Act, by which Congress had outlawed segregation of public places, applied retroactively. In theory, Wade Penny had only to move dismissal of my case when it was called for trial (the more serious assault charges having been dropped as too absurd for even an Orange County jury to believe).

Solicitor (later Judge) Thomas Cooper was not one to forgive those who had once bested him. Thus, he regularly listed the *N.C. v Klopfer* on the criminal calendar, once even having me arrested in my office at Duke when I inadvertently failed to appear in court when the docket was read. However, he never actually called the case, denying Wade an opportunity to act.

Eventually, many tedious and costly months later, the U.S. Supreme Court accepted on writ of *certiorari* Wade's contention that I was being denied my Sixth Amendment right to a speedy trial. The unanimous decision, read by Chief Justice Earl Warren, was frosting on Wade Penny's cake. Not only was this his first case before the Supreme Court, it was his first trip outside the boundaries of his native North Carolina.


## SCHOOLS AND STATIONS

When we left cosmopolitan Cambridge for the American South, it was not with any expectation that we would remain there long. But it quickly became clear that Duke had attractions. Aside from relief from chilblains, was its rural locus. Martha and I could not bear the thought of urban life. We needed space for our horses, dogs, and gardens. Riding, especially endurance racing, the grueling sport of Ride and Tie (in which teams of two people and one horse race across thirty to fifty miles of mountainous terrain), and dressage were destined to become a major outlet for our energies. The last few months at Cambridge had persuaded me that I wanted to focus on the interplay of the behavioral and ecological factors that shaped an organism's lifestyle. That required I be at an institution that could provide me with facilities close to its classrooms. Duke's campus adjoined an extensive tract of forest within which my research could easily be carried out and near which lay many old, abandoned farms. Within our first year, Martha and I also located an unkempt farm on a narrow unpaved road, whose elderly owners had despaired of ever finding a buyer. It was cheap, adjoined Duke Forest, and was only six miles from campus, on a route that could comfortably be traversed on foot, on horseback, on skis and, occasionally, by dogsled. After a few years, the road was paved, so I could also commute by bicycle or even car, though the inevitable campus parking chaos discouraged the latter. In all events, this farm became our *Tierreich* (kingdom of animals).

We were unwilling to have our children attend segregated schools and so jumped at the chance to join with a small group from the Durham and Chapel Hill Friends Meetings to found a school based on Quaker principles, and, of course, open to all. Our first class, in 1963, met in the Durham Meeting House, where carpet had to be covered by a tarpaulin to reassure some of the stodgier members that the children wouldn't damage it. From the start, it was clear we needed a separate facility, but our funds were limited. While our farm seemed out of the way at the time—today it is central to Durham-Chapel Hill—it was the only suitable site for which money was not needed. We gave a plot of land on which to build the school and thus Carolina Friends School grew out of Tierreich.

About halfway between our home and campus was a small clearing in Duke Forest suitable for aviaries and animal pens. There I constructed aviaries in which to complete the observational learning studies begun at Cambridge. I hoped that the School of Forestry would set aside a section of the forest that it managed for the establishment of a field station modeled after Thorpe's at Madingley. However, the dean of forestry assured me this was unnecessary. He would see to it my area was left undisturbed. One morning, however, in the midst of a final test series with my birds, the chatter of chain saws ended the work. On complaining to the dean of this breach of our agreement, I was told that no one in the forestry department considered cutting trees to be a disturbance. Happily, others in the administration shared a different view, and the zoology department was given a tract for its own. Thus began the Field Station for Animal Behavior Studies (now, just the Field Station), which was soon developed with funds from the National Institutes of Mental Health and the National Science Foundation. Its development brought about a close partnership with the Duke psychologists, particularly the cofounder of the Field Station, Don Adams.

Don Adams was a professor in the Department of Psychology. He was on sabbatical in Germany when I arrived at Duke, but it was clear we had much in common, despite a thirty-year age gap. Prior to coming to Duke he had been at Yale. Duke, founded in the late twenties, had called the philosopher McDougall from Harvard to head its new Department of Philosophy and Psychology. Upon McDougall's retirement, it was decided to divide the department and to bring in Adams, a well-established comparative psychologist, as chairman. Adams was initially reluctant to take the post, he later told me, because he feared difficulties with McDougall's protégé, J. B. Rhine. Rhine, originally an evangelistic preacher, had gone to Chicago to study botany. While there, he became interested in so-called extrasensory phenomena and moved to Harvard in the early 1930s, where McDougall was in the midst of studies of these and related matters (all of this according to Adams). In that first year, McDougall took off for England on a sabbatical, leaving Rhine to carry on with the studies he had begun. He trained rats in a water maze, then bred them and trained the offspring. After several generations, the rats seemed to need fewer trials to reach the training criterion than did their progenitors.

When McDougall returned from the United Kingdom, he accepted the call to Duke and took Rhine along. In due course, Rhine rose through the professional ranks. It seemed clear that as McDougall's heir apparent, Rhine would raise difficulties for whoever was brought in to replace McDougall when retirement loomed. Adams made it a condition of his coming that Rhine be removed. As Rhine had now earned tenure, this was accomplished by granting him space for an Institute for Psychical Research, which came to be financed by admirers who wanted to commune with departed loved ones.

During my Yale days, Hutchinson had introduced me to the work of Soal (1954), a philosopher-mathematician who had sought to debunk the claims of Rhine and others respecting telepathy, psychokinesis, and the like. However, he became converted to a cautious belief that there was something to be examined here, a view echoed by Hutch. I was intrigued, and became quite convinced that Soal's data, at least, were undoctored. While my suspicion was that the explanation lay in the manner in which random number lists were generated and the statistical instruments used for evaluation, this did not diminish my interest. Hence, I made it a point to attend Rhine's Monday morning coffee hours, where he reigned as paterfamilias, reading the mail to his assembled staff and asking them to report on their work. Once Robert MacArthur accompanied me. In his gentle way, he wondered aloud how the optional stopping of experiments could be reconciled with the use of particular statistical tests that assumed an a priori commitment to a fixed number of trials. "The abilities of our subjects wax and wane with their mood," Rhine explained. "There's no point continuing when they become too tired to be sensitive." Afterwards, Mrs. Rhine politely asked me not to come again.

Adams, in all events, proved prescient when he removed Rhine from psychology before coming, for it is hard to imagine how that department's record in experimental studies of behavior, a record written by such as N. Guttman (color perception), R. Erickson (neural coding), C. Erikson (mating behavior and its control in ring doves), J. Staddon (evolutionary issues), and many others, would have been established in the presence of the authoritarian Rhine.

The commonalties that established the bond between Adams and myself extended beyond our Yale heritage. He was an Elder and active member of the Durham Friends Meeting, which my family had also joined. Another intellectual bridge was built during his sabbatical, which was spent at Seewiesen, with Konrad Lorenz. Thus, when he returned to Duke, Adams was enthusiastically tuned to the problems posed by the phenomenon of imprinting, with which I too continued to be engaged (and would be for many long years after).

Finally, Adams and I found common ground in our interest in fine wines, an affection that was severely frowned upon by our Meeting (and traditional Friends in general). The consciousness of the impropriety with which our indulgence was viewed cemented our brotherhood. Our often frustrating task of designing, financing, building, and then overseeing a field station in a forest where it was unwanted and under an administration that could not care less was much lightened by an occasional postprandial tasting.

Don had one minor obsession that was unconnected to wine or science: he had always wanted a sporty Mercedes. His sabbatical at Seewiesen finally afforded him the opportunity, for in the 1950s the dollar stood high and the deutsche mark low. A car was duly purchased and, some months later, Don excused himself from our planning and grant-writing chores to fly to New York to take possession of his car. He returned sooner than expected, by air, and said nothing further of his car—nor was one to be seen. Finally, the tale emerged. As Don watched, his car was hoisted from the freighter's deck to the wharf below. Don had to stand behind a customs barricade, to which a stevedore was to drive the car. The man entered the vehicle, the engine purred, but evidently the driver misread the German instructions. The car shot into reverse and the driver shot from his seat as the car sped to the wharf's end and into the East River. Later, it developed that the shipping company would disclaim responsibility, as they had safely landed the car dockside. The stevedore company claimed it was not their man behind the wheel—he could not be identified. Don never told me the outcome of this disagreement, nor did he ever buy another car. His old wreck stayed with him till he died.

## The Duke Goat Watching Society

Martha and I had acquired our first goats, Heidi and Thor, a pair of Toggenburg yearlings, during our year as teachers at a boarding school in Western Massachusetts. This was after I had read a manuscript by Collias on socialization in goats and had decided that, along with ducks, goats

**Heidi and Thor**

would also be reasonable subjects for studies of early attachment. With a single pair, however, it was really not possible to learn the ways and manners of goats. I returned to Yale, and a year later we acquired another pair. These animals went to Liddell at Cornell when we went to Cambridge, so not until we arrived in Durham could we think about starting anew. Our first acquisition was a lovely crossbreed, mostly Nubian in appearance, whom we named Zipporah. When Southern segregationist acquaintances asked, we would delight in reminding them she was named after Moses' Black wife.

Once the Field Station was begun, facilities and support for a sizable herd became the first priority, and with NIMH support and help from nearby breeders we soon had a good-sized herd—forty or so animals—and a staff to care for them.

The first studies simply replicated in a more rigorous manner the original work by Collias (1956). We were able to show that a doe that was allowed a brief contact with one of her newborn (we arbitrarily chose a five-minute interval) would subsequently accept that kid and any others to which she had given birth even after a separation of three hours. Alien kids were rejected. Absent that five minutes of contact, all kids would be rejected. Subsequent work by various colleagues and students established other important details. We found evidence for the role of both peripheral feedback and central processes in the control of the female's acceptance of young. The hypothesis we evolved held that cervical dilation at parturition produced a central change; we postulated it was due to a release of oxytocin from the midbrain, which in turn enhanced the animal's responsiveness to olfactory stimuli emanating from its kids. There were a number of complicating details that later emerged, as well as disconcerting evidence that there might be differences between breeds, which would diminish the value of the model for mammalian systems in general. In its essentials, however, the hypothesis has proven a useful tool for exploring maternal attachment.

One of the major logistic problems the goat studies presented, aside from the obvious ones of care, feeding, and vetting (none of our local veterinarians at that time had had experience with goats or their ailments) was the necessity of our being present at the moment of birth. If the newborn could not be isolated from its mother during parturition, the animals could not be used and a full year (until the next breeding) would be lost. Initially, my family took their turns as goat watchers. All three of our daughters had become expert midwives before they reached school age. As the herd grew, it was evident that much more help was needed, even more than my group of willing graduate students could provide.

I was able to scrape together funds with which to purchase a large old three-room trailer. Along the side that featured a near-trailer-length picture window, I had erected a half-dozen birthing pens. The pens were visually isolated from one another, but, from the trailer, one had an unobstructed view into each pen. Overhead mirrors exposed otherwise hidden corners of the pens; one-way glass hid the observers in the trailer from the goats; and a simple audio-amplification system let the trailer's inhabitants hear the goats' bleats, but kept the goats undisturbed. It was thus possible for four to six persons to live comfortably in the trailer while on watch. Separate rooms provided study and sleeping facilities, and a well-stocked kitchen provided sustenance. Since all of this was but a ten-minute walk from campus, it proved a simple matter to recruit undergraduates. Most were genuinely interested in the study. Many had never seen a large mammal give birth and were interested in the process, and almost all saw life in the trailer as a welcome change from the dormitory routine. I organized seminars in which my graduate students and I explained in detail the point of our studies and trained the students in the necessary procedures. Finally, they were promised an acknowledgment in our eventual publications. Thus, with a little organization, it was possible to maintain an around-the-clock observation of all goats nearing parturition, and the experimenters could study or sleep until the final moment. Over several years hundreds of Duke students served as volunteers, many of whom later pursued careers as ethologists.

The first major article on our work appeared in *American Scientist*. "Mother Love, What Turns it On?" (1971). When I submitted the long list of names to be acknowledged, the editor balked. "Out of the question," I was told; there were simply too many names. After some discussion, a solution was proposed. The editor acknowledged that the study had been made possible through the assistance of the "Duke Goat Watching Society," a title coined for this purpose. I was given extra reprints, and my able long-suffering secretary typed individuals' names and "member DGWS" on the reprints sent to the volunteers.

Some weeks later our phone at the lab began ringing off the hook. One caller after another had questions about birthing goats, the Goat Watching Society, or both. We soon learned that a popular Hollywood TV quiz show had featured a question on goats. The authority for the answer, according to the host, was the Goat Watching Society. Evidently, the *American Scientist* article had been picked up and summarized by an Associated Press reporter, along with the unusual GWS acknowledgment. The students quickly learned of their temporary fame, so one had a T-shirt decorated with a goat face and presented this to me. It was a marvelous idea. From that day on, every volunteer received an official Goat Watching Society T-shirt. A different design and color was selected each year, and the Goat Watching Society was official.

In the years since then, I have seen GWS shirts in Stuttgart and Nairobi; friends report running into them in Sydney and Tokyo. I've no idea how many were issued during the ten or so years of the GWS's existence, but their wearers evidently are a cosmopolitan lot.

The Field Station was home to a great many different beasts. One of my graduate students pursued studies of flying squirrels there; another used it as a source for the praying mantids whose territorial behavior he was examining. Of the two ponds we built, one, in a marshy area, suited wood ducks and herons; the other, with broader, grassy banks, served a great number of undergraduate students. We also had a sizable flock of wild turkeys, a favorite species of Don Adams, and an excuse for inviting Wolfgang Schleidt to join us.

Schleidt had been one of Lorenz's earliest students, and in my opinion was the most original as well.

The work that had brought him renown in the ethological world was on a classical Lorenzian theme, namely, the nature of the releasors (or sign stimuli) that inhibited attacks by turkey hens against their chicks. These proved to be acoustic in nature, and led Schleidt to a wide range of bioacoustical problems. Adams wanted Schleidt to come help him with a general investigation of the behavior of the (rare) wild turkey. Most so-called wild birds are game-farm stock whose behavior has been much modified. As Adams once explained, if you mimic a turkey's call, the game farm birds stretch their heads up, while the wild ones duck. As a wild-turkey hunter, Adams claimed he wore out more shells putting them into and then removing them from his firearm than he ever discharged.

I had other reasons for wanting Schleidt to come. One of my students was engaged in an ambitious study of mockingbird song. First, he wanted to compare the repertoires of North Carolina mockingbirds reared in artificial environments with and without birdsong present. Then he wanted to collect mockingbirds in the Galapagos, where the species had spent many generations in an environment with only a small variety of songs (in contrast to that in North Carolina), and repeat the study with them. Did the ontogenetic conditions outweigh the phylogenetic constraints? In the event, the experiment could not be fully completed. However, it looked promising at the time, except for a number of practical difficulties, not the least of which was my modest technical skills vis-à-vis bioacoustics. In the early sixties this was not the highly developed field it is today. Manuals were scarce and equipment had largely to be self-assembled. Schleidt would be

**A fallow deer** *(Dama dama)* **at the field station**

just the person to educate us, and, happily, he agreed to come. (I would have been even happier if he'd never left.)

The most conspicuous of the station beasts were doubtless our herd of fallow deer. For our studies of socialization in goats we needed a species suitable for cross-fostering. Deer recommended themselves because in some ways the mother-young relation complemented that in goats; while kids tended to take the initiative in nursing, and would follow their does till they were fed, fawns lay quietly until the doe took the initiative and roused them. I went hunting for deer and found them at the New York Zoological Garden. The zoo knew what they were doing when they shipped a herd of *Dama dama* (fallow deer) to me; I didn't. Fallow deer seemed to me advantageous as they were more nearly goat size than our native species of white-tails, and this would also allow me to use lower fencing. This was no trivial detail, as I had to enclose about forty acres of woodland for the herd. As I soon learned, however, keeping the deer within the enclosure was the least of our problems. Fallow deer, unless hand raised and socialized from birth, tend to be about as intractable, nervous, and generally obnoxious a research animal as one could imagine. The bucks were edgy and aggressive, and we were fortunate to suffer only injuries and no deaths among our students in the years we held them. The cross-fostering never did succeed

Much of the hand rearing (for we went through several seasons of this in order to establish a herd made up of animals that had never had experience with adults of their kind) was done by my family. All three girls became experts at mimicking the doe's call to nurse and, along with our usual pack of elkhounds, we had fawns underfoot in every room from March to June.

The explosive growth of the herd, and the lack of interest on the part of zoos to take animals off of our hands, left us with no alternative but to kill off the annual surplus, which by fall was often as many as twenty animals. Field Station students and staff grew very tired of venison in those days. For years after, when a venison roast was retrieved from our big freezer for the sake of special guests our girls would groan despairingly, "Venison? Again?!" Even though it was not stylish in those years we all had leather clothes. Indeed, I found several dozen frozen hides still awaiting tanning in the departmental freezers in '92.

My student Barrie Gilbert's studies revealed, among other things, that the hand-reared animals would easily form a structured herd, and used the same motor patterns as did the controls when it came to communicating. However, the "meanings" of particular signals often did not seem to correspond. For example, young hand-reared stags moved their heads in a way that, among normally reared males, signaled attack, but in the case of the hand-reared animals it was a friendly sign.

Barrie went on from studying deer to work with Gregory Bateson, first in John Lilly's lab in the Virgin Isles, then in Hawaii, using dolphins and whales as subjects for studies in social processes and communication. When Barrie went to Utah (Logan) to join the faculty, he had become one of the most experienced field workers of his generation. Thus, when he focused his attention on grizzly bears we were hopeful that we would soon have answers to any number of the questions they posed. Unfortunately, a nearly fatal encounter with a sow put an early end to Barrie's study. I had heard a radio report of a bear attack on a ranger, but had no idea Barrie was the victim until, months later, I received a letter that began laconically, "As you can see, I am still alive. . . ."

## THE PRIMATE CENTER

Much of what would appear to be "classical" ethological work on primates is today conducted by anthropologists. Indeed, probably as many members of the Animal Behavior Society or the International Ethological Conferences consider themselves anthropologists as comparative psychologists.

The merging of anthropology and ethology, which began in the mid-sixties, was in part due to some of the leaders shifting their interests from birds or rodents to primates and then attracting students from anthropology to their labs. In part it was also due to the tireless fieldwork of anthropologists such as Ray Carpenter. However, another major contribution to the merger, who himself actually never did any behavioral work, was the colorful and prolific anthropologist John Buettner-Janusch. He became an example of another way in which interdisciplinary networks develop.

"B-J," together with his wife Vina, was examining the structure of the hemoglobins of the primates, and in particular that of the prosimians. To this end, and in order to gain insight into patterns of hemoglobin heritability and variability, he had assembled a colony of lemurs and galagos, along with an odd baboon and chimp or two at Yale, where he taught. Yale was not keen to allow him to develop a full-scale primate colony on its urban campus, so after viewing the expanse of the Duke Behavior Station, he proposed a collaboration: if I'd arrange space for his animals at the station, he'd make them available for behavioral studies. He proposed that our long-term goal ought be the development of a major research facility that would jointly serve our common interests.

This idea was highly attractive to anyone interested in behavioral comparisons. The prosimians were a "tightly-bound" group, phyletically, but were adapted to a range of different ecological conditions and had adopted various lifestyles, some solitary, others highly social. B-J's collection was the world's largest, and, in fact, once grew to almost a thousand individuals of over twenty species. Our pact was quickly signed in 1965, and within months B-J's beasts moved into a facility he referred to as The Crummy Old Goat Barn. In fact, it was officially The Camel Barn. Though designed for goats, the invoices for its construction came due just prior to the activation date of the grant that was to pay for it. To avoid embarrassment, I asked a colleague for help. He had unexpended funds available that had been designated for a shelter for his camels (which were still in Australia, where, in fact, they ultimately remained), and with these funds he covered the construction costs.

In the months that followed, when B-J and I saw how fruitful our collaboration could be, we enlisted an imaginative architect to help design the facility. B-J merely needed breeding facilities. I, however, wanted space in which the animals could lead more-or-less natural lives. B-J had to periodically catch the animals in order to draw blood. I wanted to be able to watch them unobserved. The compromise was a series of hexagonal rooms (round ones would have been preferable, for the absence of corners results in the animals treating their confined space nearly as if they were unconfined).

**The Crummy Old Goat Barn**

However, local builders were unable to deal with such structures, so the architect devised the hexagons. They were two-storied, with overhead and ground-level observation ports with one-way glass, and were connected to outdoor enclosures by means of electrically powered doors. The arrangement proved ideal for both of our groups and the only (though considerable) friction came from occasional changes in the blood-drawing or light-dark schedules.

B-J was easily the most colorful figure on the Duke campus. His dress varied from highly formal three-piece suits to Nehru jackets and sandals. The former were adorned with incandescent neckties, the latter embellished with equally conspicuous beads or plumes. No public event or controversy could pass without a colorful, often unprintable, epithet from B-J. Nor were there dinner or party invitations that were more coveted than those from Vina and B-J, for they were masters at entertainment. Even a spontaneous, "drop by for supper tonight" meant, at the least, gourmet frankfurters, potatoes, sauerkraut, and the finest champagne.

After some years at Duke, B-J left to become a department chairman at New York University, where a bizarre series of events ended his career. Vina died during exploratory surgery, and B-J began conspicuously consuming drugs, principally quaaludes. He was arrested, convicted, and, probably due largely to his unrepentant and aggressive court demeanor, sentenced to several years imprisonment.

B-J spent his initial term at Eglin Detention Center in Florida. Very soon after his release he was rearrested on the more serious charge of having mailed poisoned chocolates to the judge who had sentenced him (as well as to others). He was then sentenced to forty years in the grim maximum security institutions of the federal penal system. His many letters from within provide a chilling picture of how our penitentiaries, originally conceived by their Quaker founders as places for meditation and penitence, have degenerated into medieval dungeons. B-J died in prison in 1992. The primate colony, now the Duke University Primate Center, with a quasi-autonomous status, has continued to thrive and will doubtless outlive both its founders.

The early behavioral studies at the Primate Center were conducted by Duke graduate students, along with scores of undergraduate honors students. However, it was not long before students from other departments and institutions began arriving and imparting an international, interdisciplinary flavor. Matt Cartmill, from Duke's Department of Anthropology and Anatomy, brought morphological studies to the center, while Carl Erickson from psychology added a behavioral-developmental emphasis. Elwyn Simons, recruited to be the director of the Primate Center, and his group discovered a potential for paleontological work with primates, and Ken Glander, the current director, has added an ecological dimension. Indeed, the center has become a miniuniversity, but with the difference that all its diverse members share a common interest in prosimian primates. No other center with a large collection of prosimian primates, and certainly very few primate centers of any sort, can provide their animals with enclosures consisting of several hectares of natural forest, as does Duke. This has made the center a unique facility.

# 5

## Flights to the Field

### Here and There

Tinbergen and Lorenz repeatedly stressed the importance of knowing the animal in its own environment. Studies of confined or hand-reared animals were a major occupation for them, but always against the background of familiarity with the unfettered beast. My own more limited experience had certainly reinforced that view. At Yale, I had the chance to contrast the behavior of lab-reared ducklings with those hatched at Dillon Ripley's Litchfield estate. Later, when I completed my doctoral research at the Delta Waterfowl Station (in Manitoba), I noted that even under the relatively free conditions of the Litchfield Preserve, there were differences from Delta.

The Delta experience, my first fieldwork as a "professional" biologist, also brought me to the realization that it afforded the best possible opportunity to get to know other biologists. I've since spent hundreds of hours visiting colleagues in their labs, meeting with them at conferences, and even carousing after hours, but never felt I could establish as firm or honest a rapport as when we met, tired, wet, mosquito-wracked, and hungry, at the meal that marked day's end in the field. I can still, for example, recall the conversations, the details of the experiments and their rationale, and some of their personal quirks. Some of the several dozen graduate students I've had are but faint ghosts. I have to reread their theses to recall who they were, what they did, and why, and this is true even in cases in which I worked long and closely with that student. However, in those cases in which we had been on an expedition together, even if for but a few days, that person's ideas and personality are an indelible record, for better or for worse. Thus, I've become a proponent of fieldwork for biologists for reasons in addition to those very important ones the old naturalists enunciated.

Much of my fieldwork has been in the Caribbean basin and Central America. Sometimes by design, usually by accident, this allowed me and my students to establish contacts that might otherwise not have come into existence. For example, when Barrie Gilbert and I needed to chase banana-quits, a small bird *(Coereba flaveola),* from island to island (we were inter-ested in discovering if the breadth of their niche varied with species diver-sity) we were sent to Archie Carr for logistical advice. Carr, a consummate naturalist and specialist on sea turtles, had up to that time had little contact with ethology. I doubt Barrie and I were of any real help to him, but that fieldwork established a bridge that was later used by Archie's students and mine. Another trip, involving the same quest for bananaquits, brought an-other student (Jack Hailman) and me to the Smithsonian Station on Barro Colorado Island. This led to our coming to know Martin Moynihan, BCI's irascible director, and, important for Jack, a leading authority on gulls. Other old friends happened to be on the island, too—Robert MacArthur and Monte Lloyd—so Jack had an opportunity to meet them, not to men-tion the legendary Ted Schneirla.

Long the patron of animal behavior at the American Museum, Schneirla's influence came principally through the work of his students and colleagues—foremost, perhaps, Danny Lehrman, but also Ethel Tobach, Lester Aronson, and Jay Rosenblatt. Schneirla's style of writing was so turgid and laden with jargon that few people I know (outside of friends in psychology, who seem to enjoy arcane argumentation) read his work. In fact, Schneirla spoke as he wrote. Like Pogo's friends, the words emerged in Gothic script: "I have tintinnabulations tonight," he reported once, to explain why he was not joining a conversation.

The bananaquits led me and Ken Crowell to Costa Rica, where we made contact with a group of visiting American biologists who were form-ing a consortium for the study of tropical biology and a facility for training biologists. This became the Organization for Tropical Studies, on whose board I sat for a number of years early in its history and which has since become a major force promoting studies of and in the New World tropics. OTS is a facility that has particularly promoted the interaction of students from North, Central, and South America in ecology, behavior, and conser-vation.

Sometimes fieldwork includes tedious technical chores. On one trip we needed to ascertain the selectivity with which birds fed. This we hoped to do by collecting all the seeds available to our seed-eating species in randomly selected plots and then to compare them to the seeds recovered from the stomachs of a few individuals we captured. Differences in the proportion of seed types in and outside the birds' guts would provide a measure of selectivity. Identifying and counting seeds is a boring, time-consuming business, so we filled a suitcase with vials of seed for later analysis at home. We had to fly home through Miami, which meant an encounter with customs agents suspicious of our scruffy, too-long-in-the-tent appearance. We were more than casually interrogated and searched. "What's this?," was the puzzled query when the agent came to our case of vials. Weary of the delay and wanting to minimize discussion, I said succinctly, "Nothing to interest you, just grass seeds." "Grass?," was the startled reply, and the red lights on the counter began to flash. We were quickly surrounded by armed agents, but their chief, noting my consternation, abruptly turned and left. "Anyone so square as to claim he's hauling grass can be trusted!," he laughingly shouted as he retreated.

Fieldwork, of course, often meant extended absences from home. Whenever possible, I tried to take my family along, or at least some part of it. Given our extensive menagerie, which required we find reliable, animal-wise house sitters, there were always complications. These were not always those anticipated. Once, a student and his young wife took charge of our household. While Ron finished the barn chores, Jan prepared an elegant supper and set a proper table—their student quarters hadn't previously allowed for such civilities. The food ready, but Ron not yet in, Janice stepped out to the barn to help him finish chores. On her return, the neatly laid-out table was as it had been left, but the dishes were empty and clean. Assuming this to be a prank by one of their classmates, she and Ron waited patiently for the laughter. When hunger overcame patience, a make-do meal was created. Only on our return, when this tale was told to us, was the mystery solved. One of our less scrupulous elkhounds had the habit of jumping onto tables and carefully cleaning off the platters, displacing no more than the lid to the sugar bowl (he had a sweet tooth).

Occasionally fieldwork at a particular site or on a particular problem was carried on by several generations of students. This had the great advantage of allowing fresh sets of eyes to review the scene. For example, Dan Rubenstein, as a sideline to his thesis, initiated a study of the social dynamics of the feral ponies on the offshore islands near Duke's Marine Lab. The ponies on one end of the largest island were territorial; the dominant males maintained harems on a specific bit of ground, while on the other end they did not. This was evidently related to the availability and defensibility of water holes, but only later did this become apparent. Dan watched the horses for several years until his principal study was completed; then his place was taken by Rolf Hoffmann, who had come to me from Schmidt-Koenig's lab in Tübingen. After Hoffmann came Beth Franke-Stevens, who still continues to keep an eye on "her" ponies (besides fulfilling her responsibilities as an associate director of the Atlanta Zoo). Each generation of pony-students thus had the opportunity to elaborate or modify the constructions of their now well-established predecessors and then defend their interpretations directly. It's a great way to learn, for the predecessors, however famous, are not necessarily more technically adept at watching ponies.

Some of our fieldwork was on sites too small or remote to allow much contact with other scientists. These, at least, had the advantage of fostering close and often enduring friendships. I think, for instance, of our studies of the aggressive behavior of pomacentrid fish, coral-dwelling species of tropical waters. Some of these fish allow members of one species to approach more closely than those of another. We thought to use this behavior to determine whether changes in the threshold of responses to different species varied independently. The traditional views of Lorenzians would predict they would not. For the purpose of this study, the Smithsonian Institution allowed us the use of an islet off the Belizean coast that they'd leased. It was probably three-quarters of an acre in extent, so the crew of Martha, me, daughter Erika, a handful of students, and a local lady, whom we knew only as Regina and who proved a royal cook left no space for others.

Bill Kier was just commencing his studies of cuttlefish locomotion, and was the most competent marine biologist among us. His sense of the constraints exercised by physical mechanisms in locomotion was a useful antidote to the tendency others of us had to regard the behavioral possibilities as limited only by neural events.

Purely solitary fieldwork has played a minor role in the lives of most ethologists, though the outstanding work by Jane Goodall on chimpanzees certainly belies any claim that this work hasn't been of major significance. However, both the expense of getting into the field and the psychological problems of isolation do limit opportunities.

Two of my field studies are worth noting here because of the logistical and legal problems they raised, by no means unique to these studies but with the potential for producing particularly unwanted results. One of these involved work in Madagascar, the other in California.

The island of Madagascar, once connected to the east coast of southern Africa, is home to a variety of endemic species, of which the lemurs are probably best known. I had been studying maternal behavior in our captive colony and needed to confirm my results in the field. Several graduate students had ideas of their own they wanted to explore, so we planned a series of trips, the work being expected to span four to five years. The first trip involved just myself and Jonathan Harrington and had us located at a private preserve in southwest Madagascar, at Berenty. Alison Jolly née Bishop, who had also worked under Hutchinson at Yale, had based her marvelous book (*Lemur Behaviour,* 1966) on her studies at Berenty, so I knew it to be a suitable locale. Indeed, one happy circumstance of using her study site was that I was able to remap the territories of certain of the animals she had identified. It was this circumstance that led to our first meeting and production of a short paper with her on the long-term stability of lemurs' territorial boundaries.

Jonathan and I had a busy time watching and tracking our animals. To avoid alarming them, we learned to station ourselves in their midst before daybreak, so we would already be present when they woke. We would not depart until after dark, when they had settled down for the night . Our own domestic chores had therefore to be done by the dim light of a kerosene lantern. The plan was to have one of us prepare supper while the other cleaned the dishes from the previous day's meal—a time-sparing arrangement. However, it seemed as if the washup was never necessary—the dishes were always clean, even if not neatly stacked. We attributed this to one of the natives who lived nearby, and resolved to leave him a useful gift on our departure.

Nighttimes were always noisy, for our rude hut was within the forest. One night there was such a clamor from the kitchen hut that we grabbed flashlights and went for a look. What we saw were our dish cleaners at work: a bevy of giant (10–14 cm long) Madagascar cockroaches, fighting and hissing at each other over our dishes, which were being tossed lightly around, like so many Frisbees. We did wash up after supper thereafter.

Our trip was altogether a success. Jonathan produced a thesis, I confirmed lab results, and a long, friendly relationship with Jolly was begun, and later reinforced when she became one of the external advisors at the Duke Primate Center. The ground for the entry of the others was laid, too. Norm Budnitz produced a much-cited study, but then he discovered teaching at Carolina Friends School was even more fun than research. His scholarly work continued, but this was to be his last trip to Madagascar, as it also proved to be for Cathy Jones Dainis. Lee McGeorge, on the other hand, caught the attention of Gerard Durrell when the latter was visiting the Duke Primate Center. He wanted to establish a bioacoustics lab at his Isle of Jersey preserve, and as Lee's study had been an analysis of the relation between sounds produced and propagation characteristics of the forest, she seemed an appropriate candidate for the post. In fact, she proved more than just that, for the two married a couple of years later and Lee's subsequent publications covered many more topics than lemur vocalizations.

In the meantime, physical anthropologists had also discovered Madagascar and were making regular visits, particularly John Buettner-Janusch and his associates and students. Among these was a respected scientist who strolled a beach one sunny day in the company of a German acquaintance.

The German suggested a swim, an offer that was declined along with the observation that the coastal waters were infested with sharks. The German thereupon swam by himself, and some days later bits of his corpse were found on the shore. The local police evidently found it hard to believe anyone would freely venture into the surf, so the more prudent of the pair found himself being detained on suspicion of murder. Criminal charges against Westerners in Third World countries cannot be taken lightly, so it is easy to understand why the suspect seized the first opportunity to obtain a different passport and slip away. The resulting furor effectively forbade a return of U.S. citizens to Madagascar for several years; by then funds were in scarce supply, so our studies were never completed. A political revolution in Madagascar at the same time brought into power a pro-Maoist, anti-U.S. regime, and that might equally well have been the cause for denial of our visas.

The California trip produced problems of my own making, but it may serve as a warning to others. I had been looking for another species with which to test certain predictions that came from our goat model for understanding the attachment of newly parturient animals to their offspring. Publications by my favorite teacher at UCLA, George Bartholomew, suggested that the elephant seal might be just the beast, and, with his help, I established the contacts and permits to work at Año Nuevo Island, an abandoned Coast Guard station off the California coast, near Santa Cruz, and San Nicholas Island, one of the Channel Islands farther south and a U.S. Naval base. The students accompanying me were to be Barrie Gilbert and Amelie Segrè. Our local scientific contact was to be a young behavioral scientist, but he tragically died as we were in the midst of preparing a joint grant proposal.

Our intent in this study lay first in describing the normal course of interactions postpartum, then to examine the responses of anosmic (incapable of smelling) mothers to their own and to alien young. In the course of our work with goats, we had established that the most reliable means of inducing a short-term anosmia with a predictable recovery rate was by administering a nasal spray of 10 percent cocaine in HCl. The idea for using cocaine was provided by Knut Schmidt-Nielsen, the eminent physiologist whose lab lay across the hall from mine. (The close and lasting rapport between physiologists such as Amiram Shkolnik, Barrie Pinchot, Richard Taylor, and the animal behavior group was largely due to our propinquity, along with the Schmidt-Nielsen midday tea.) The various other topical anesthetics I'd tried all had varying time courses from one trial to the next, except for cocaine. Schmidt-Nielsen had learned of this from his experience with the resistance during the Second World War. According to his account, when German troops searched Danish vessels seeking allied pilots who'd crashed or other wanted persons, they and their search dogs would march up gangplanks between two rows of straight-backed fishermen whose hands would be outstretched in a pacifying gesture as the dogs passed. On the backs of their hands they had dusted cocaine powder, which the dogs inhaled with every sniff. Even without the ever-present fish, none of these dogs would ever find a hidden person.

We'd constructed a half-meter-long syringe with which we hoped to apply the cocaine, and, if that failed, we also had a syringe projector, a rifle that used compressed $CO_2$ to shoot a syringe. Impact would prompt the discharge of a small gunpowder cartridge, driving home a tranquilizer, immobilizing agent, or anything else one might wish to administer.

Cocaine, of course, was a controlled substance, so I had to apply for a license to use it. We filed an application and over the course of the next week received a daily phone call. Each was from a higher station in the governmental chain of command. The question was always the same: Was I using the drug for elephants or for seals? In those days, the sixties, the now prevailing hysteria regarding drugs did not exist. The puzzlement as to my purpose was considerable, however.

Once the permit was issued, I discovered the five hundred grams to which I was entitled were not available through legal channels, the only ones I could afford, so I had to settle for about half that quantity and scale down our sample. Then I had to apply for a permit to use the drug in California, a different jurisdiction. Some kind soul agreed to be the official user, so, permits in hand, we prepared to board our flight. The morning of our departure, however, I belatedly discovered that permits to buy and use the drug in North Carolina and California were not enough: yet another permit was required to transport the stuff over state lines. No time was left for legal niceties: we transferred the powder into a jar labeled "powdered aspirin," resealed it, and were on our way. Our one moment of panic came when we were challenged by a security guard as we were boarding. Wasn't that a gun case we were carrying? "Oh," I innocently replied, "this is just a syringe projector." That was okay with him, though later, racing through Chicago's airport to make a connection, gun in Barrie's hands, we were sure a policeman would stop us. By then, fortunately, the cocaine was safely checked, so a search would have produced nothing more incriminating than our immobilizing drugs. Thinking back now of the potential consequences of my casual attitude towards regulations does give me chills.

I've alluded to one of the problems of working with captive animals, which is that one can never be sure that the behavior being observed is not an artifact of captivity. To guard against being fooled, I've always made it a rule to do parallel observations in the field. Even here there can be problems, however, and not merely stemming from drugs. As mentioned, we had begun our study on Año Nuevo Island, off the Northern Californian coast, at the site of an abandoned Coast Guard station. Hazardous tides and currents made bringing in supplies a nightmare, exactly the reason the Coast Guard abandoned the place, so we moved south to one of the Channel Islands, San Nicholas. With the help of the head of the Office of Naval Research, we were permitted onto the island, a naval base, and comfortably housed in the officer's barracks. Of course, I was considered a security risk because of my felony conviction from UCLA days, so I could not be allowed to roam freely around the base. (Many years later, the California Supreme Court ruled, in a case similar to mine, that when "imposition of sentence is set aside" the effect is to nullify the conviction. I am a felon no longer.) A young ensign was assigned to me, who had the thankless job of shadowing my every move. In a jeep specially commandeered for the purpose, he would drive me the few meters from barracks to mess hall to library, and, of course, the beach where the elephant seals lay. During the hours we spent seal watching, our driver practiced jumping his jeep across crevices in the hard, sandy cliffs, until he misjudged and lodged the vehicle upside down and some meters deep. Thereafter we got to walk.

The seals, who were busily calving, seemed to pay us no heed. We all but leaned against them while writing up our notes. Not so the patrolling harem masters. When the 1.5-ton bulls approached, we had to retreat quickly. Fortunately, their great size limited the distances of their rushes, and they had to cool off for some minutes before making another rush at rival or intruder. I gauged the time for my retreats closely, but on one occasion the laces of my boot on one foot were caught in the hooks of the other boot. Down I went, with the shadow of the behemoth looming above me. Only Barrie's quick distraction maneuvers saved me from ending like a fly under a swatter.

The females paid us no heed, though they were otherwise quarrelsome, and often bit calves, though usually not their own. This was what we'd expected, but Barrie was still skeptical and insisted we repeat observations from a distant dune using a telescope. His intuition was correct: the quarrels ceased, and alien infants were readily nursed.

For many years our principal research beasts had been goats, though only the domestic variety. We finally had a chance to compare their feral

**Elephant seals at Año Nuevo Island. Photos by Barrie Gilbert.**

cousins when Dillon Ripley, then secretary of the Smithsonian Institution, called us. The British Royal Society had a research station on Aldabra Atoll. The Society was concerned about the depredations of the goats.

It was at just this time that Meg Gould had decided to join the pony watchers on the Outer Banks, but with the thought of studying the feral goats, which shared the islands. The low, thick, live oak forests in the center of the larger islands greatly limited visual contacts between animals. How would this influence herd structure? Our observations of gazelles in the deserts of Israel had already convinced us of the fragility of herd structure. Meg was soon off for Aldabra, there to discover a treasure trove of unexpected insights. I didn't want to leave my family for the one to two years of Meg's stay, so we planned for me to join her midway through her stay, and to try to plan the trip so as to catch the twice-yearly supply ship on its return from Mombassa to the Seychelles, shortening our stay to, we hoped, about ten weeks. For that period of time, Martha and I decided our girls could fend for themselves along with the help of their grandmother and my delightful and talented research assistant.

## JOURNAL FROM ALDABRA

Aldabra is a fly speck in the Indian Ocean some four hundred kilometers northwest of the giant island of Madagascar. The atoll resembles a flattened doughnut, thirty kilometers long, its width varying from about five to ten kilometers. Most of the interior, the doughnut's hole, is a shallow lagoon that connects to the sea through three passes that dissect the narrow rim of the doughnut and divide it into four major islands.

We flew to Mahé, principal island of the Seychelles, early in December, boarding a crowded, cramped flight that took us from L.A. to London, and then Nairobi, before finally disgorging us. Because we had boarded directly upon both of us completing our best-ever runs in the Culver City marathon, our limbs felt more than normally cramped and sore. But we'd arrived, and the next day our ship would sail. All we needed now was to find clothes to replace those in our lost luggage.

In affirming the importance of fieldwork, Tinbergen also urged keeping a detailed journal, a practice I had found easy to avoid: after an arduous day recording data in uncomfortable situations, sitting over a stack of blank pages with no explicit instructions on what to record was not attractive. Aldabra was different from earlier field sites, however. Our subjects were only visible during short periods at the end of the day, we were free of our usual routines, and we were largely isolated from others, to say nothing of mail, phone, or radio. No laptops, either. With lots of time to dwell on the many new scenes before us, beginning aboard the ship that took us to our island, Martha and I commenced a field journal. To provide a flavor of biological studies in exotic places, excerpts from our journal are provided in the appendix.

# 6

## The Later Years

### BEHAVIORAL ECOLOGY

I tend to identify myself as a behavioral ecologist, a term that acknowledges my early training as an ecologist under G. E. Hutchinson as well as my later ethological studies with W. H. Thorpe. Indeed, the first book I wrote I entitled *Behavioral Aspects of Ecology* (1962).* In the early sixties this was an idiosyncratic designation, but times changed and the term is now commonplace.

The discipline we know as behavioral ecology represents the intersection of ecological and behavioral studies, a hybrid of these two fields, ecology and ethology, with questions not asked by its "parents". How do species share space and use habitats? Why don't predators overeat their prey? What keeps species distinct? It's generally accepted that behavioral ecology's beginning was marked by the publication of Crook and Gartlan's 1966 study of the relationship between environmental conditions and social structure in primates. It has, however, occurred to me that one could equally well argue that behavioral ecology was antecedent and gave rise to ethology (ecology having arisen independently in the 1920s, cf. Hagen 1992).

Obviously, chronological continuities don't necessarily translate to causal connections, but in this case both continuities can be demonstrated. Von Uexkuell's *Umwelt und Innenwelt der Tiere* (1909), literally the "outer and inner worlds of animals" (usually translated as the "perceptual world of animals"), is clearly an example of a behavioral ecology study, and it influenced both Lorenz and Tinbergen, as judged by their references to Von

---

* The book was actually conceived together with Robert MacArthur, but midway through our collaboration we decided to develop separate volumes. His tome on island biogeography has become a classic.

Uexkuell in various of their writings. Both of them, in turn, had a direct impact upon the many early papers by Crook (1960) on social behavior in weaver finches.

Crook was a student of W. H. Thorpe at Cambridge in the mid-fifties, just at the height of the postwar interest in European ethology. Crook's descriptions of the quality of nesting materials as assessed by his birds is vividly reminiscent of Von Uexkuell, who, though his work is not cited in Crook's thesis, was known to Crook. (I recall conversations on Von Uexkuell during my Cambridge days, conservations that took place in Crook's presence. And, as may be noted from the acknowledgments in Crook's papers, Tinbergen was of no small importance to him. Tinbergen, in turn, frequently referred to Von Uexkuell.) O. Heinroth, on whom Lorenz lavishes unqualified praise for his inspiration and intellectual leadership, also drew upon Von Uexkuell's work.

The early history of behavioral ecology is inextricably entwined with the political developments in Germany. Von Uexkuell and Speman probably helped shape Lorenz's view both on the strength of their science and the congeniality to him of their politics. A personal friendship between Lorenz and Tinbergen bridged political differences, and allowed Tinbergen to assimilate some of his colleagues' theoretical baggage, especially the viewpoints of Von Uexkuell. Both Tinbergen and Lorenz had considerable influence upon Cambridge's respected W. H. Thorpe (Thorpe 1979). Students who came to Thorpe almost always were exposed as well to Tinbergen—J. H. Crook, as noted, is a striking example of this. Ironically, Thorpe despised the German political scene as much or more than did Tinbergen, so behavioral ecology was, after all, freed from any particular political associations.

Meantime, at the University of Chicago, a group of investigations associated with Warder Clyde Allee independently developed a line of inquiry that also led directly to the behavioral ecology of the 1960s and beyond.

Like Thorpe, Allee was a member of the Religious Society of Friends who took his Quaker beliefs seriously and seems to have allowed them to influence his choice of research problems. Much of his life's work was focused upon the mechanisms of development of animal societies and the important role of cooperation in evolution. His views and his studies have been reviewed in detail by Mitman (1992), who calls attention to Allee's close association with the embryologists W. Patten, E. Conklin, and D. S. Jordan, who championed the role of the biologist as a prophet of a new social order. They opposed Social Darwinism, favored eugenics, and saw cooperative principles as dominating natural systems. Allee's study of animal aggregations, in the light of his own and his colleagues' philosophic predilections, could not but lead to Allee's magnum opus, a study of cooperation in animal life and evolution (Allee 1938).

Allee received his doctorate at Chicago just before World War I under the supervision of V. Shelford, a founder of modern ecology. At this time, ecology was firmly embedded within physiology (autecology, as opposed to the study of populations or communities, which came a decade later, and was briefly known as synecology.) Allee issued first his studies of aggregation, one of the first major enterprises in behavioral ecology, and then his work on cooperation and group selection. The latter was undoubtedly inspired in part by another Chicago biologist, the geneticist Sewall Wright. Allee also collaborated with Wright's brother, Quincy, providing him with biological insights for his monumental study of the history of war, which seems to strengthen the view that Allee was less successful than he claimed in separating beliefs and science.

Allee's studies of animal societies were to play an important role in the development of the ideas of V. C. Wynne-Edwards, who came to champion behavior and group selection as the proximate and ultimate causes, respectively, for population regulation. This placed him at odds with the influential David Lack, the Oxford ornithologist. Lack developed the notion of "optimum" clutch or litter size, suggesting that too many offspring could imperil parents' fitness as much as too few. This idea of optimality was further developed mathematically by Robert MacArthur, and in the hands of subsequent generations of behavioral ecologists became a major orienting principle in the behavioral ecology of the 1970s.

Meantime, W. H. Hamilton's seminal papers on altruism provided a basis for reconciling the competing claims of group selection and traditional Darwinist selection. "Fitness," Hamilton argued, is ultimately a function of the proportion of an individual's genes that are passed on. Since siblings share genes, cooperation or altruistic acts directed towards relations need not reduce "inclusive" fitness.

The notion of inclusive fitness became a mainstay of behavioral ecological thinking, especially as its theoretical (mathematical) implications were developed by Maynard-Smith. Then, in the hands of E. O. Wilson, R. Trivers, and their students, it was formed into an important element of sociobiology. Wilson believed sociobiology would ultimately absorb behavioral ecology (along with ethology and a number of related fields); instead, of course, issues of social evolution have become part and parcel of the behavioral ecology agenda.

Two concepts appear to have become the dominant organizing principles: first, optimality, which was introduced through Lack, MacArthur, and, by implication, Crook, Thorpe, and Tinbergen; and secondly, game theory, which grew from the work of W. D. Hamilton, G. Williams, Maynard-Smith, and indirectly, E. O. Wilson and R. Trivers. All of these workers were influenced by (if only by their reactions to) Wynne-Edwards, who, we have noted, built onto the constructions of Allee, whose scientific beginnings can be traced to the early years of biology at the University of Chicago under Whitman (Mitman 1992). Whitman had some slight influence on some Europeans, notably Lorenz.

Behavioral ecology, which today represents the most active domain in which ethologists (in the broad sense) operate, thus appears to have been formed by the union of two distinctive lineages, one traced forward from Von Uexkuell and his largely European following, the other from Allee and his Chicago compatriots. The irony is that the "parents," Von Uexkuell and Allee, were both socially engaged and had philosophic political beliefs that seem clearly to have influenced their science, though those beliefs could scarcely have been more dissimilar.

## SOCIOBIOLOGY

The story of behavioral ecology cannot be concluded without some mention of E. O. Wilson's efforts to absorb it within a broader field that was to include ethology, much of ecology, and cultural anthropology as well: sociobiology. Sociobiology's popularity today seems to be greatest in domains far removed from biological study, especially the social sciences. The hard core of the biological work that might be judged sociobiological (i.e., relevant to understanding the evolution of cultures) is firmly rooted in behavioral ecology. Shortly after the publication of E. O. Wilson's *Sociobiology* in 1975, however, it was not self-evident that this would continue.

His goal, in his own words, was "to codify sociobiology into a branch of evolutionary biology and particularly modern population biology. When the same parameters and quantitative theory are used to analyze both termite colonies and troops of rhesus macaques, we have a unified science of sociobiology" (1975, 4).

*Sociobiology* concluded with speculations on the evolutionary origins of human cultural activities and their genetic control or basis. At worst, these thoughts could be criticized as naive; at best they were ideas that, though not novel, could be reevaluated in the light of some new evolutionary concepts—specifically, kin selection. Wilson's thoughts on the subject were clearly labeled speculation, not dogma.

I think many of the readers of *Sociobiology* were as surprised as was I when a savage review of the book appeared in the popular press, authored by R. C. Lewontin, population geneticist and professor, first at Chicago, then at Harvard. The review went far beyond the claims actually made by Wilson, and represented a philippic against the notion that genes somehow control human behavior, a view later cogently and eloquently developed in "Not In Our Genes" (Rose, Lewontin, and Kamin 1984). While I had long shared Lewontin's position on the issue of genetic determinism, I hadn't viewed Wilson's book as a particularly serious contribution to that particular debate. Lewontin's response to Wilson's book was pragmatic: genetic determinism had become the favored theory of the political right; it seemed to justify labeling (and discounting) all sorts of social misfits, criminals, welfare recipients, gays—in short, anyone outside the proper pall. The contradictions and absurdities in this view had been thoroughly aired, but Lewontin, ever the evangelist, wanted to make the point again and before the broadest possible audience. An attack on the renowned Wilson would bring Lewontin's views the desired publicity, even if the target, Wilson, was somewhat misrepresented.

Lewontin had been a political activist from his earliest professional years. His early activities were efforts to improve the quality of high school science education. During the 1960s Civil Rights struggle in North Carolina, when a number of us (Duke professors) were jailed, Lewontin was quick to come to our support. During the Vietnam War, he resigned in protest from the National Academy of Science, whose committees he perceived to be assisting the war effort. His activities opposing right-wing manipulation of scientific concepts were consistent with a long-held philosophy that related beliefs and actions.

Wilson, by contrast, could almost have been considered reclusive. I cannot recall a single conversation on political issues in which he participated. His interests seemed wholly focused on his scientific work, which was as prodigious in quantity as it was outstanding in quality. I cannot but believe that Wilson was caught totally off guard by the vehemence of the attack against him, which was soon taken up by a Cambridge-based, largely student-run group, Science for the People. Their attacks were not limited to reviews or editorial tirades, but included picket lines outside his lab and lecture hall, and even an assault during a public scientific talk.

Regrettably, few of us within the biological community reacted quickly enough. George Barlow, an ethologist at U.C. Berkeley, initiated an effort to depersonalize and depoliticize the dispute, but responses were slow in coming. I too wrote a defense of Wilson (though insofar as the scientific issues were concerned, I agreed with Lewontin), but the usual peer review process delayed the appearance of my article for a full year. Those who quickly rallied behind Wilson were to a large degree drawn from the fringes of the scientific community; in many cases they were persons with political axes to grind no less massive than Lewontin's and with fewer scruples about how to wield them.

Clearly, what I state here is a personal opinion, but from my correspondence with the principals, I came to believe that the orchestrated attacks on Wilson, figurative as well as literal, the support by right-wing interests, and the lack of support from his scientific peers, politicized Wilson. The books he subsequently produced, in any event, were often unabashedly political, shaping scientific observations and theories so as to conform to a scheme established a priori, just exactly the sin for which he had been so unjustly castigated originally.

The belief that science is or ought to be value-free still is held by many, though it has been increasingly challenged in this century. Some, such as the evolutionary geneticist Theodosius Dobzhansky, have promoted a form of scientific evangelism, in which science replaces religious and political ideals. Dobzhansky explicitly promoted a "new humanism" derived from evolutionary theory (Dobzhansky 1967). Harvard's B. F. Skinner did much the same, if more stridently, in *Beyond Freedom and Dignity* (1971).

Others, particularly in this decade, claim that science ought to be (or in effect, is) controlled by society at large and not left to its devotees. Thus Feyerabend (1978), it seems, would make conclusions on scientific issues subject to a vote.

While Wilson probably would align himself with Dobzhansky, his work has been characterized as claiming "that the present states of human society are the specific results of biological forces and the biological nature of the human species. . . . These theories have operated as legitimizers for such institutions as war, male dominance of females . . ." (G. Allen et al. 1975). Yet Wilson denies any intent or desire that his evolutionary explanations serve as political justifications. The fact that (Wilson might say) our genes incline males to dominate females is no reason that we cannot alter our behavior (and our laws) to preclude male dominance. Ironically, Lorenz made the same claim vis-à-vis his arguments regarding the effects of race mixing, though, unlike Wilson, who has been consistent throughout, Lorenz's disclaimer came years later.

The lesson of the sociobiology debate seems to be that the scientist's stance on the relation of his or her work to society's problems is irrelevant. Whether a purest or a pragmatist, the fruits of a scientist's labors are harvested (or not) according to criteria independent of that scientist's desires. It does not seem to me at all likely that the debate would have erupted, or had such an influence on the body politic, had Wilson's initially innocent remarks not been exploited by advocates for a particular political platform.

His own ivory tower was pulled down around him. This may have been due to a larger necessity: while the tower protects its inhabitants, it is not conducive to the development of a larger vision that integrates science and society. The alternative is, however, equally problematic: as technical problems come to dominate social issues, science comes ever more to be blamed for those problems and seen as a threatening ideology. "Know-nothing" movements, modern equivalents to the Luddite movement of the Industrial Revolution, are increasingly common.

Somehow there needs to be fashioned a rationale by which society can exploit scientific developments. This rationale must take heed of the holistic character of major scientific developments. One cannot select a twig and exclude the tree.

## BEHAVIORAL MECHANISMS

The answer I like best to the query, What is ethology?, is the charge given in Tinbergen's questions: What is the function and history of each behavior pattern? How does it develop? What is the mechanism? Early in ethology's own history it was the first query that received primary emphasis. In the past three decades, it has become the study of mechanisms. This has brought ethologists into close alliance with neurobiologists, with, I believe, benefits on both sides. While neurobiologists are often (and justifiably) contemptuous of the uncritical fashion in which models of neural mechanisms are developed and tested by the "classical" ethologists, the rigid reductionism of the former has been no less criticized. At the 1992 Seville conference on neurobiology and behavior some neurobiologists suggested that social attachments in young primates are regulated at least in part by biogenic amines. This was one of a growing series of efforts to link complex behavior of whole organisms to particular sites, cells, or chemicals in the central nervous system.

But how compelling is the evidence that links structure and functions in general? In neurobiology, the view that particular neural areas or chemicals serve specific functions has been prominent since the days of Gall, Broca, and Wernicke. Work of the past several decades has largely strengthened their premises. Indeed, few procedures in biology have been as straightforward and elegant (technical difficulties aside) as those that entail recording the consequences of stimulating particular neurons or ganglia, or interfering with or augmenting the flow of particular neurotransmitters. The bulk of our information on the mechanisms of central processing has come from studies utilizing such procedures. But have they told us "the whole truth"? The burden of my argument is that they neither have nor can do so, not because of methodological shortcomings, but because the interpretations of the results are premised on an erroneous view of the fixity of structure-function relationship, as Lashley suggested years ago.

The notion of a stable structure-function relationship is solidly embedded in the biological corpus. It has a long history and is expressed in a variety of ways, particularly in post-Darwinian evolutionary biology, which, in turn, has had a pivotal influence on all of modern biology.

But if we critically review the prevailing orthodoxies respecting the evolution and control of behavior, we often find that the experimental evidence that is claimed to support the traditional views is not compelling— that other conclusions are equally reasonable and equally compatible with the data.

An example of a structure-function relationship that is often cited is the tale of increasing industrialization of the nineteenth century that led to atmospheric pollution and, thus, to tree trunks becoming darker. Moths that alighted on these trunks were more conspicuous to avian predators if the trunks were light rather than dark. Kettlewell (1961) was able to document a shift in the proportions of a melanic (dark) phase of the moth *Biston betularia* that tracked this trend. Further, he experimentally evaluated the hypothesis. In an unpolluted area, where tree trunks were light, 6.3 percent of captured and released melanic moths were later recaptured, compared to 12.5 percent of the light moths. In a polluted area, however, with dark trunks, the proportions were 27.5 percent melanics to 13.1 percent light moths, a dramatic reversal. Kettlewell also verified that birds did more often prey upon the more conspicuous lighter moths rather than melanics when these sat upon dark trunks. As trees darken, it seems a darker race of moths evolves. This has become a textbook example.

Jack Hailman (1982), however, has pointed out that there is actually no evidence that *Biston betularia* consists of a single interbreeding, polymorphic population rather than two separate populations of different colors, whose numbers vary independently. Further, moth densities of up to $10^5$ per km raise questions about the significance of the relatively few moths taken by birds. Hence, Hailman argues that it does not necessarily follow that the rise of melanic moths was due to selection against their lighter brethren, though this has been heretofore universally assumed. The light and dark moths could be members of different populations whose numbers vary independently. *Biston betularia* is but one example, of course, and the Modern Synthesis depends on a great many others (Huxley 1942). However, if one examines the evidence for the three pillars on which the Synthesis rests, the appropriateness of a metaphor by B. L. Slobodkin, an ecologist of note as well as a Talmudic scholar, may be appreciated (1951). He'd compared research to ritual prayers. In each case, the procedure—research and prayer—merely prepares the mind for a revelation unrelated to the procedure itself.

This is of direct relevance to the structure-function issue in neurobiology.

Consider first efforts to understand the organization and function of the CNS through comparative studies. These are premised upon the principle of homology. This principle states that similarities in structure due to a common ancestry are sufficiently conservative as to allow their use in establishing phyletic relationships. Konrad Lorenz in his Nobel address, epitomizes its use: "A great part of my life's work has consisted in tracing the phylogeny of behavior by disentangling the effects of homology and of parallel evolution. Full recognition of the fact that behavior patterns can be hereditary and species-specific to the point of being homologizable was impeded by resistance from certain schools of thought. . . ." (p. 231). One of those who resisted these impediments was Gregory Bateson, who countered: "There is a certain elegance in the notion that evolutionary process must generate or differentiate two types of comparability: analogy, generated by the active process itself, and homology, generated by the failure of that process to change its own production. Clearly, this logical elegance would be spoiled by any other form of comparability which might demand recognition" (Bateson 1960). Bateson then goes on to enumerate the immense difficulties in making the distinction, especially among related organisms where it is the matter of the resemblance that is adaptive, not the structure itself. Alternatively, resemblances that appear to be adaptive responses to selection, analogies, may represent physical constraints that are only fortuitously adaptive—as is possibly the case for the similarity in wing patterns of edible butterflies and their inedible or toxic "models."

The rejection of the homology-analogy dichotomy as a guide to disentangling phylogenies, and especially phylogenies of behavior, at once precludes extrapolation from one species to the next. The decorticate cat cannot then be seen as having its reptilian brain revealed, and the study of central function must proceed de novo with each species.

The important role of social context in the expression of behavioral mechanisms is emphasized in a study by Miller et al. (1990). They knew, from earlier work, that mallard ducklings become immobile when exposed to a particular maternal vocalization (the "alarm" call). They also knew that this behavior was contingent upon prenatal auditory experience: devocalized, socially isolated ducklings are generally unresponsive to the alarm call, though exposure to the call will, after a few repetitions, reestablish the response. However, the experiments that led to these conclusions were performed with individually reared and individually tested ducklings. Rearing and testing devocalized ducklings in a social group resulted in no loss of the freezing response to the alarm call, even though these ducklings were as lacking in auditory experience as their isolated counterparts. The experimenters concluded that auditory experience is necessary in some contexts but not in others—that there exist multiple pathways in behavioral development. There are other possible explanations of their results as well.

Yet another case that must trouble cortical mappers is Lorber's (1980) work on cortical representation in hydrocephalics. One particular subject he describes is a normally functioning graduate student in mathematics who, as a result of infant hydrocephaly, has a cerebral volume of not more than 500 cc and only a few millimeters of cortex. Clearly, conventional views on localization of function only apply here if ontogeny is ignored, or at least only with the qualification that reassignments are possible in ontogeny. Hooper and Moulins (1989) show that individual neurons can switch from one functional network to another as the consequence of sensory-induced changes in membrane properties. In their system, the functional membership of a neuron in a central network is not even fixed ontogenetically. The conclusion I draw is that whether we consider nervous systems as small as that of *Aplysia* or as complex as that of *Homo sapiens,* knowing a neuron's name and address doesn't assure knowledge of its function.

The Seville participants' (cf. p. 44) descriptions of the behaving organism posed no particular problems to me. Neither did their description of ontogenetic changes in brain cytoarchitecture and the distribution of bioamines. Implicit in the descriptions is the assumption of a causal relation, and it is this feature which is inconsistent with the argument I make above. To put it more bluntly, the effort to explain phenomena at one level of organization (behavior) by reference to selected phenomena at a lower level (chemical events) represents a type of reductionism that has not yet assisted us in understanding the complexities of animal behavior, nor does it offer much promise.

What I think is a critical difference between work such as that of Dethier or Roeder and that of the Seville group and many others is this: the former had precisely characterized the behavior whose mechanisms they were analyzing; they knew how it varied from occasion to occasion as well as between individuals. Heinroth, Lorenz, Tinbergen, ethology's founding parents, have always insisted that detailed familiarity with the species be a prerequisite to all behavioral work, but it is often lacking among neurobiologists.

I think very few of the students from Duke succumbed to this sort of thinking, although it's also true that only a few focused on mechanisms, per se. Hailman, in his study of the factors that led gull chicks to peck more avidly at some model beaks than others, found a mechanistic explanation for the chick's color preferences. He showed that the different droplets in the retinal cells differentially absorbed light of varying wavelengths. By assuming that one class of cells acted to stimulate, the other to inhibit, a response, he was able to model the color choices of the chick.

Hailman's "Ontogeny of an Instinct," the title of his thesis, was a marvelous example of an integrated response to Tinbergen's questions. Ironically, when Hailman first appeared in my lab—he was merely accompanying his wife, a Duke alumna, who wanted to visit old friends and familiar places—his interest then seemed to be solely in the descriptive aspects of behavior.

We got along so well on the occasion of that first meeting that the Hailmans moved their camping rig from a nearby state park to our front yard so that we could continue our conversation. A year or so later, they came to Duke and through interactions with the Schmidt-Nielsen group and the biophysicist Don Fluke (who shared an interest in the physics of vision) Hailman developed his thesis.

Not surprisingly, Hailman's work quickly attracted attention and was a major factor in the creation of further ties between several groups of investigators. First was Schwartzkopf's lab at Tübingen, Germany, in 1985. I had introduced the two to one another after having gotten to know Schwartzkopf at an ethological conference in Starnberg, Germany. He had spent an afternoon with me on a tour of old churches in the region, a subject on which he was an expert. Unfortunately, he had lost one leg and wore a prosthetic that squeaked with each step, a disconcerting effect inside a silent Gothic chapel.

Schwartzkopf at that time was in von Frisch's lab and had no independent post until he was later called to the chair at Tübingen. This elevation to power evidently affected his personality, because he soon became an autocrat of the traditional Teutonic type. Students and associates were not even permitted to open the lab's mailbox in his absence; he had to sort the letters and then dole them out. Once, in Schwartzkopf's absence, the mailbox bulging, Hailman forced it open to claim his mail. None of the others in the lab would accept theirs, however. His association with Tübingen ended prematurely, and Hailman went on to Lehrman's Institute at Rutgers.

Lehrman's "Institute for Animal Behavior" was for two decades one of the two or three most productive and exciting centers for ethological studies that focused on mechanisms. My first personal contact with Lehrman came about when he visited Don Adams at Duke and was invited to sit in on Gilbert Gottlieb's thesis defense. Aside from our common scientific interests we quickly discovered a common love for early-morning birding—Lehrman was phenomenally good at field identification of warblers, which he had learned, he said, in Central Park while truant from school. I found it hard to believe he could be as much the city boy he claimed to be and still be so good a naturalist. I became convinced he was a city boy, however, when I later visited him at his Greenwich Village apartment. We were enjoying sherry before an open fire when it became time to return to the institute for an evening lecture. As any country boy would, I began to bank the blazing logs with ash. Lehrman pushed me aside, seized one log after another with a pair of tongs and disappeared with them into a back room. When he took the last of the several logs, I followed and found him dunking them in a bathtub full of water, then placing them against a warm radiator. In response to my quizzical look, he explained each log cost several dollars plus an arduous trip up several flights of stairs, no small task. On Lehrman's next trip to Duke, I had ready a cord of well-seasoned hickory logs with which we overloaded his car—an old tank of a Buick. He'd decided he could persuade his students to carry the wood up his stairway.

Few other studies of mechanisms were carried out by my students, despite the stimulus provided by visits of Lehrman and his group. The oxytocin–maternal care work with goats has already been described. The most notable other effort was a simulation study of perception by Rob Gendron, which built close ties to John Staddon in the psychology department. Eventually, in fact, Staddon became an adjunct member of the zoology department. Another psychologist, Carl Erickson, further assured a continuing exchange between those students interested in mechanisms, for Carl was a brilliant protégé of Lehrman's. This network of contacts and interrelationships still continues to increase in complexity.

Gendron, whose work with Staddon more or less formalized the psychology-zoology connection, worked with marine organisms, carpenter bees, and a host of other creatures before devoting himself to computer-driven studies of perception. These were an outgrowth of his thesis problem, in which he was reexamining L. Tinbergen's notion of the "search image." When predators search for cryptic prey they may suddenly demonstrate an increased rate of discovery, as if they had developed a clearer image of the desired prey. This could be due to a change in search speed, however, as well as to the formation of a search image. Using both quail and food pellets that were dyed to be more or less conspicuous, and human subjects selecting "prey" on a computer screen, he was able to demonstrate the operation of both strategies. His work has nicely corroborated more naturalistic studies of this phenomenon, notably of Oxford's versatile behavioral ecologist, Alex Kacelnik. Kacelnik appeared late on the behavioral scene due to the decades of totalitarian rule in his native Argentina, a delay for which he has compensated with a prolific output of original papers on prey detection and feeding behavior. Not all political interference can be so readily compensated. G. Tembrock was isolated in East Berlin when the wall cut off the DDR from the rest of the world. His failure to make concessions to the ruling powers prevented his work from attracting the attention it deserved. He was not allowed to travel, and his correspondence was limited. Had he lived in West Berlin, Tembrock might well have been celebrated as one of the founders of bioacoustics and behavioral ecology. As it is, he is scarcely known outside of what was East Germany.

When Gendron came to Duke he had already completed studies of marine organisms, so I was not surprised to learn he was an experienced SCUBA diver. This was precisely at the time that I'd returned from Aldabra resolved to study aggression with pomacentrid fish as my subjects. To do this I first had to learn to swim—no small feat for someone with a lifetime fear of deep water ("deep" being anything above my nose, an attitude probably attributable to several episodes of near drowning in childhood). I overcame the fear by hiring my animal caretaker, Sue Bergeron, as swim instructor. An expert swimmer herself, a self-styled "Army brat" with the demeanor of a Marine drill sergeant, she inspired more fear in me than did the water, so under her tutelage I swam. Diving lessons followed, and for my first expedition to the lair of the pomacentrids (in the Florida Keys) I chose Rob as my diving buddy.

One of our dives was made into a deep well within a coral mass that happened to be lined with a fine calcareous sediment. On descending, my flippers stirred up the sediment and, in seconds, we were engulfed in a thick white fog. Orientation was impossible, and I panicked. Rob, obviously anticipating my reaction, firmly grabbed my shoulder and held me in place until I'd calmed and the water cleared, allowing a slow, safe ascent. Had he not restrained me, I'd surely have popped to the surface like a cork, my lungs exploding in the process.

### NETWORKS AND OTHER CONNECTIONS

A short history of the study of animal behavior in North America has been written by Donald Dewsbury (1989). In his article he graphically portrays the overlapping tenures of individuals interested in animal behavior at the University of Chicago, Harvard, and Johns Hopkins. An even more vivid indication of the close contacts between individual ethologists is given by is table 1 (below), which depicts the North American participants in the early days of the International Ethological Conference. Many of the participants had had frequent and close contacts with one another prior to the meetings, and generally even more after. This assured the establishment of an informal network of active ethologists who kept closer tab of one another's work and students through private correspondence and intervisitation than through the established journals. Since only slightly more than two dozen Americans were present at the 1965 Zurich Conference, it is apparent that such cohesion was easy to achieve: if the younger participants weren't students of the older ones (which was generally the case), they were at least well acquainted. Thus, individual quirks of character and personality could have a disproportionate influence as compared with the present-day Ethological Conferences or Animal Behavior Society meetings, whose participants number many hundreds. Among such hordes, personalities must now be extraordinary to stand out. Does this mean that trends, new ideas, or novel paradigms are less easily initiated or brought to acceptance?

In any case, the familiarity that members of the ethological community had with one another in the "good old days" was born of small labs, frequent small conferences, and considerable intervisitation, to an extent impossible to achieve today. The experiences I had in my first decade at Duke were characteristic of the period and demonstrated the origins and reinforcements of the community network.

Lunchtime in the zoology department at Duke in the late fifties meant gathering for a strong brew of tea in the laboratories of the Schmidt-Nielsens, Knut and Bodil, both distinguished physiologists (she was the daughter of Nobelist August Krogh, he was Krogh's student). Lunch was a lively, informal affair; there was an unspoken rule that we would avoid those subjects that related to the fate of round objects that were kicked, thrown, struck, or otherwise abused or moved about a field or court. The usual attendees included biophysicist Don Fluke. Our association led to his assisting me in tracking the movements of rodents in a habitat selection study. I had placed voles in large pens that offered three types of habitat and wanted to monitor the time the animals spent in each section. Don proposed injecting radio-cesium into the animals and forming a grid with dental X-ray film on the floor of the cage. The density of the developed film would be proportional to the time the animals spent near it. This study was aborted when Wecker published his meticulous study (Wecker 1963), which answered the same questions I'd asked, but Don's involvement in this bit of behavioral work soon led to this becoming an invaluable ally of generations of students lacking training in physics.

Table 1. North American Ethologists at the
International Conference (from Dewsbury 1989)

| Zurich (1965) | The Hague (1963) | Starnberg (1961) | Cambridge (1959) | Freiburg (1957) | Groningen (1955) |
|---|---|---|---|---|---|
| Aronson | Aronson | Beach | Aronson | Beach | Beach |
| Brown | Banks | Beckwith | Bullock | Beckwith | Davenport |
| Cooper | Barlow | Dethier | Crane | Dethier | Hess |
| Denenberg | Beach | Dilger | Davenport | Hess | Lehrman |
| Eisenberg | Bitterman | Fuller | Dethier | Lehrman | |
| Fentress | Brody | Hale | Dilger | Roeder | |
| Ficken | Collias | Harlow | Emlen | | |
| Flynn | Dane | Hess | Hess | | |
| Freedman | Eisenberg | Klinghammer | Lehrman | | |
| Gottlieb | Emlen | Klopfer | Moynihan | | |
| Griffin | Ficken | Kramer | Pittendrigh | | |
| Kleiman | Gardner | Lehrman | Schein | | |
| Klopfer | Ginsburg | Levine | Scott | | |
| Lehrman | Hess | Moltz | Stokes | | |
| Levine | Komisaruk | McCleary | Verplanck | | |
| Mason | Konishi | Reese | | | |
| Maynard | Lehrman | Roeder | | | |
| Myrberg | Maynard | Schein | | | |
| Payne | McConnell | Shaw | | | |
| Phillips | Nelson | Winn | | | |
| Reese | Payne | | | | |
| Rosenblatt | Roeder | | | | |
| Scott | Rosenblatt | | | | |
| Shaw | Ward | | | | |
| Warren | Wilson | | | | |
| Wilson | | | | | |

Dan Livingstone was another lunchtime regular, when he wasn't dodging crocodiles on the lakes of Africa. On one expedition, a large crocodile bit through both chambers of the rubber raft on which Dan and a student were traveling. They had to abandon ship and swim a considerable distance to the croc-infested shore. Their adventure first came to light when I received a card from Africa in which Dan referred offhandedly to having survived. I immediately phoned his wife, Bert, also a zoologist and former Hutchinson student, but the earlier written letters of explanation had not reached her. It was an anxious week for her before the full tale was told.

The Schmidt-Nielsens always had postdoctoral associates from abroad with them, and many of them also formed lasting connections with the Behavior Group. Steven Thesleff, from Sweden, became interested in modeling neuronal mechanisms in memory (Thesleff 1962), and thus led to my first trip to Sweden and a meeting with Eric Fabricius, one of the earliest students of avian imprinting.

The arrival of Amiram Shkolnik from Israel brought about an even more complex web of relationships. Shkolnik, homesick for his family, spent considerable time in our home, becoming particularly attached to our ten-year-old, Lisa, who was the same age as one of his daughters. Thus, on my next trip to Madagascar what was more natural than that I should drop Lisa off in Tel Aviv to spend a few weeks with Amiram's daughter at Kibbutz Kabri in the lovely hills of the northern Galilee? On completing work in Madagascar, I joined them and came to know the enthusiastic group of ethologists and ecologists around H. Mendelsohn, the patriarch of the group and a specialist in reptilian behavior and ecology. Yoram Yom Tov, Amotz Zahavi, Eviator Nevo, and Elan Golani have since become well-established investigators, lecturing to audiences around the world. However, when I returned to Tel Aviv for a semester's teaching in 1970, several of them were still just graduate students, and we shared the excitement of having found kindred interests between such distant labs. That small group, two generations later, has expanded its contacts and relationships so that there is today scarcely a center for behavioral or ecological study group anywhere between Antwerp and Zanzibar without an Israeli connection.

I've already alluded to the Australian connection through Andrewartha, which was reinforced through the subsequent arrival of several postdoctoral scholars, among them the eminent Peter Bentley, who remained in North Carolina for many years as a professor at N.C.U.'s School of Veterinary Medicine. Subsequently, an invitation from A. J. Marshall, professor at the newly established University of Monash, nearly led me to emigrate, but when I declined, Klaus Immelman, a German ethologist, went instead.

Another visitor to the luncheon group introduced yet further dimen-

sions to the network. C. R. (Dick) Taylor, fresh from Kenya, joined K
Schmidt-Nielsen for two years before continuing on to Harvard to form his
own locomotion-energetics lab there, a happy blend of behavioral, mor-
phometric, and physiological analyses. When Taylor first went to Africa to
study the energetics of locomotion his funds ran out long before he had the
results he needed. To obtain his data he had had to tame and train his animals, cheetah and elephants included. He discovered that his subjects could
provide for their own care through fees paid by U.S. television companies
for their use in filming. Years later, when viewing such shows as *Daktari*
Taylor would exclaim, "Hey, that's my . . ." Dick's interest in locomotion
was probably not unrelated to the fact that he had been a track star as a
schoolboy. He continued to run while at Duke, and our countless miles
together on road and track are the basis of all my knowledge of the energet
ics of locomotion.

The other dimension was provided by Jürgen Aschoff, a visitor from
the Max Planck Institut at Andechs. The Institut's fame, however, was due
not only to the work on circadian rhythms that Aschoff pioneered but to its
proximity to Kloster Andechs, an ancient monastery renowned for the quality
and potency of its brew. The combination of Andechs beer and Aschoff's
talent for organizing parties gave his institute a unique status in Germany
Aschoff brought both his intellectual talents and ebullience with him, and
life in the Schmidt-Nielsen lab was never again as lively after he left.

Aschoff's work on circadian (twenty-four-hour) rhythms linked closely
with the major connector between the local scene and German science
Klaus Schmidt-Koenig. Klaus had originally come to work with Gaithe
Pratt, but then joined my lab and built a substantial research group around
the problem of pigeon (and other) orientation. By the time Klaus left Duke
a decade or so later, he'd achieved considerable renown and succeeded
Hans Peters as professor at the Behavior Institut at Tübingen.

On an early visit to Klaus at Tübingen, I was introduced to the tall
distinguished, very formal Herr Dr. Professor Peters. Suddenly I had a rec
ollection of a scene enacted decades earlier at a party at Lorenz's institute
a busload of us, midnight revelers, drove to Starnberger See for a midnigh
swim, au naturel. Lorenz mischievously hid the clothing of one of his guest:
so that later, while everyone else was settled clothed in their seats, "De
schöne Peters" was rushing up and down the beach, bathed in our headlamps
seeking his clothes. It was hard for me not to giggle when we were intro
duced.

I subsequently spent a year (1979–80) as Schmidt-Koenig's guest
thanks to the award of an Alexander von Humboldt Preis, which led to an
extensive exchange of our respective students. It is unlikely that there is a

Prof. Klaus Schmidt-Koenig of Tübingen, as he is today (more or less), and as he was when he first came to Duke in the late 1950s

"Der Schöne Prof. Peters," in Tübingen, 1980

PK (nearest center) at a reception by President Carstens (to his left) of the Feder:
Republic of Germany in honor of awardees of the Humboldt Society, September 197

major German university that does not have a more or less direct connection to that early (1960s to '80s) Duke group, a pattern that doubtless mirrors what occurred in a number of other institutions.

International conferences are intended to help develop networks by bringing together established and newly minted investigators. In the early days of ethology, when the numbers of its notaries was small, this did happen. It was at one such, for instance, that I met Pat Bateson, now provost of Kings College, Cambridge, and cousin to Gregory, who provided links to yet another universe. Pat and I, building on that first of many meetings, eventually spent twenty years as coeditors of the series *Perspectives in Ethology*. Most international meetings, however, were too large or formal to do more than provide a stage on which the famous could perform or venues in which old friends could be found. The exceptions were the "working conference" meetings that focused upon a particular theme and whose participants were individually selected and invited. Among the most successful of these in my experience were those organized by the Wenner-Gren Foundation and held at their former conference center, Burg Wartenstein, in Glognitz, Austria.

Margaret Mead, who had made something of a study of the designing of conferences, believed the ambiance of the conferences to be vitally important to their success. Comfort was not, she believed, essential to creativity and a successful outcome, but departure from the routine was. Meetings could be held in mud-floored, thatched huts or in the most outrageous luxury; ordinary classrooms or board rooms, however, would not do. Wenner-Gren provided for one sort of ideal locale. Glognitz was the last mountain outpost before the plains of Vienna, and Burg Wartenstein served as a protective fortress from about the twelfth century or so. It was gloriously perched, hundreds of feet above the small town below. It appeared to have sprung directly from the rocks that served as its foundation.

The restoration of the Burg had not done away with its breath of antiquity. Nor were the concessions to modernity conspicuous. Lights and light switches were cunningly concealed; central heat was evenly distributed, but its source invisible. In the spacious dining hall, refilled platters appeared magically when needed. (It was only after my third visit that I discovered hidden buzzers used to signal the kitchen.) Discussions were recorded and transcribed, but the intimidating electronics could not be discerned. Needless to say, radios, newspapers, casual guests, and visitors were rigorously excluded. Except for a single night on the town in distant Vienna, conference participants at Burg Wartenstein were isolated with their colleagues for the full seven to ten days of their work. The esoteric environment, as

Mead anticipated, contributed in no small part to the magical syntheses that took place there.

**The book *Our Own Metaphor*, by C. M. Bateson, identifies this group posed by the Wenner-Gren's Burg Wartenstein Conference Center. C.M.B. is in the center; her father, Gregory Bateson, is third from the right, next to Anatal Holt, next to PK.**

To be sure, not all successful conferences—success itself has several measures—need be as ecumenical as were those at the Burg. Later that same year, on another occasion, a meeting at the Burg brought together a contingent of leading anthropologists and zoologists with interests in primate communication. This more homogeneous assembly was also effective in developing collaborative and long-term relations. Such affairs, I suggest, have been a significant factor in the development of ethology in the 1950s and beyond.

# 7

## Reflecting on My Trade

### TEACHING

The immediacy with which animal behavior studies can be appreciated, their seemingly obvious (if often spurious) application to human conditions, and the charms of the subjects themselves place ethologists in an enviable position. They have a far more direct access to the lay public or to beginning students than do their colleagues in more esoteric fields. Perhaps this is why so many introductory biology texts have been written by ethologists, or by biologists with major interests in behavior. The breadth of the subject matter with which ethologists need be familiar may also contribute.

My own efforts in this direction, a book written with a physiologist colleague, Marvin Bernstein, and superbly, if abstractly, illustrated by Martha Wittels Szerlip, was never published. The few reviewers who liked it were outvoted by the many more who didn't. I myself rather liked the book, for it was based upon an introductory biology course that Bernstein and I had offered at Duke.

This course, admittedly, was unorthodox, but it honestly reflected the way Bernstein and I (and many of our ethological colleagues) thought about science, rather than the way traditional texts stated we should think. Since I was not obligated to teach—I had been given a Career Development Award by the NIMH, which allowed Duke to hire another faculty member to assume my duties—we decided to offer our course (in the late 1960s) as an alternative option to the regular Intro. Bio. Students were not compelled to take our course, so we decided we could use our own criteria for grades. In place of the usual letter grades, students were told that only one grade would be given, A, but not until work of A quality was presented. Several possibilities were offered: an outstanding performance on an examination, a term paper on an approved theme, or an independent research project, and we decided to impose no time limit, nor refuse repeated reexamination. So long as our standards were met, we reasoned, it didn't matter how long or how many trials it took. Wasn't that how our own research operated? Students who failed to complete the work were simply not acknowledged as having ever taken the course. We counted on small numbers to overcome the anticipated administrative problems.

I continue to believe our conception was sound, and someday I hope to test it again. However, our efforts failed dismally. Many students simply wouldn't believe we were serious about A or nothing and complained about being denied a right to a "gentleman's C." The registrar took umbrage at delays in the reporting of grades and filed a complaint with Duke's University Faculty Council, (the equivalent of a faculty House of Representatives). A major squall erupted when this body voted to censure me for violating university rules on the reporting of grades, without prior notification that this was on their agenda. I learned of their action after the fact when a reporter from the student paper phoned me for a comment. My response was unpremeditated, intemperate, and rude. Our attempts to mold introductory biology after our research ended abruptly when the departmental chairman took over the course. I can't yet decide whether there's a moral to this tale.

By the 1980s, the teaching of animal behavior brought with it political problems, as the animal rights movement gathered strength. Introductory courses dropped the use of living vertebrates from their labs and physiologists also found it easier to use simulations or invertebrates than to cope with bureaucratic procedures. Those options were not feasible for ethologists, nor were we off the hook when we limited ourselves to merely observational studies; extremists in the movement declared it wrong to hold animals in captivity at all. I learned what this meant one morning when I arrived at the lab to find my chick pens broken open and the remnants of my flock scattered about the parking lot and nearby woods. The neighborhood dogs (obviously also uncaged and free) had had a feast.

I must confess to a certain sympathy for the animal rights movement. Years earlier, I had been asked if I'd consider testifying before a congressional subcommittee in opposition to pending legislation that would regulate the use of research animals. That legislation seemed unnecessary to me, as the animals in Duke's zoology department were all well cared for. It also seemed to me that there was far greater need for regulation in the meat, pet, and farm sector than in science: the proposed legislation smelled like a not-so-subtle attack of antirationalists against the scientific establishment (I continue to suspect this). In all events, to prepare myself for an eventual trip to Washington, I made the rounds of labs in other areas of the university. My shock at what I found forced me to become a supporter of federal regulations.

At the Medical Center, for example, in the 1950s and early '60s, unclaimed dogs were brought in each week's end from surrounding pounds. They were turned loose in a small tower room atop the Medical School, and, if they arrived late on a Friday, spent the weekend without sufficient food or drink. When I arrived on a Monday morning, the attendants were engaged in removing the mangled bodies of the nonsurvivors. Humane treatment aside, one wonders how reliable studies could be that were carried out on such a motley, ill-cared-for pack.

The issue of appropriate and humane treatment is not the only thing that the animal rights movement is about, however. Two of its early leaders, Tom Regan and Peter Singer, for example, oppose altogether the exploitation of animals, whether for food, as pets, and certainly as subjects for scientific research. With a growing recognition that animals are not machinelike automatons, a view that Don Griffin has done much to make respectable even in psychology labs, these concerns have ceased to be trivial. Whatever the motivations of the zealots, biologists in general, and ethologists in particular, are now compelled to address the issue: Do animals have rights?

Paradoxically, in medieval times, in Europe, animals did enjoy judicial rights, as in the instance of a formal trial for alleged misdeeds (Evans 1906). When rats "wantonly" and "feloniously" destroyed the barley crop of the Autun province in the sixteenth century, they were ordered to court and professional advocates were appointed to plead their case. In this instance, charges were not pursued because local cats made it impossible for the rats to safely make their way to court. A pig that ate a child was less fortunate and was condemned to death by being burned alive. Other times, pigs who had attacked children were ceremoniously hung. Animals were even placed on the rack. Not that confessions were expected, but this was part of the legal procedures of the time. Animals were accountable to the law and also given legal rights.

The Enlightenment, and especially Cartesian mechanistics, altered this, and by the late nineteenth century animals no longer were legal equivalents of persons. Instead, laws began to appear requiring those who had or used animals to do so humanely. In large part this grew out of Kantian views that our treatment of animals reflects itself in our treatment of other people. Following Darwin, animals were regarded as hierarchically arrayed, with species near the top of the evolutionary tree deserving more consideration than those below. Curiously, in the United Kingdom, which has long regulated the use of animals for research, horses, dogs, and cats were ranked above primates; a researcher might more easily gain permission to vivisect a monkey than a dog. (These rules do not, I believe, apply today.)

"Humane care" legislation, however, imposes an obligation on humans. It does not grant that animals have rights nor that they may themselves or through their representatives apply to court.

The past two decades have seen efforts in the U.S. courts to allow such applications. In the *Mineral King Valley* case, which went before the U.S. Supreme Court, the Sierra Club attempted to become "guardian" of the valley so as to represent it and protect it against destruction. The novelty was that the club did not argue that its own interests were at stake (Stone 1972).

The case was lost, but the Sierra Club's attorneys have pointed out that there are many precedents of institutions and even ships being treated in the courts as persons. Why not valleys, trees, and, above all, animals (Stone 1987)?

The particular contribution of ethologists may lie in working towards a clarification of the notion of "consciousness." If animals can be regarded as sharing this quintessential human attribute, the claim that they deserve more regard than ships and corporations would inevitably be strengthened. This would then allow for new and different challenges to the *Mineral King* decision, but the day has not yet come. In the meantime, we must contend with activists who would kill animals to "save" them, and regulations that mandate climate controls for animal rooms, including those holding only aquaria.

"Animal rights" is not the only arena that ethologists have to enter. There is also that arena we know as Social Darwinism. When E. O. Wilson published his heavyweight volume, *Sociobiology,* the initial reaction was one of awe that a single biologist could amass such a wealth of material and discuss it in such an enlightened, literate manner. Wilson had made his debut on the scientific stage as an expert on ants and their systematics. I first got to know him in the sixties when he came to Duke for a field foray and borrowed one of my student assistants, Henry Hespenheide, as a helper. Later, Wilson collaborated with my old friend Robert MacArthur to develop a theory of biogeography. Even at that time, Wilson evinced little special interest in the behavior of vertebrate organisms and their social evolution.

Admittedly, much of our time together was not spent discussing science but rather running. Ed was a frustrated trackster, frequently trying but never quite succeeding in running a sub-six-minute mile. Since I'd some background in coaching track (for some years I coached the Duke women's club), he looked to me for training advice. A few years later, he was induced to travel to Duke to accept an honorary doctorate. The ceremony fell on the same weekend as the Southeastern Masters Track Championships, where a subsix mile seemed possible, and evidently as important as Duke's degree. Alas, a sprained ankle put him out of contention, but he elected to come anyway to cheer me on and to watch Martha set a North Carolina woman's age record in the mile. (She later set the U.S. record.) Ed's absence at the various precommencement gatherings was attributed to his desire to confer with scientific colleagues. An alert reporter spilled the beans, however, and I later got an earful from an irate Terry Sanford, then president of Duke, for our frivolities.

Lest it be thought that racing together inevitably bonds biologists, I should mention an egregious exception. It began with a phone call from Dan Rubenstein at Princeton inviting me to address their ecology group. It had, however, been years since I'd published in this field, so why the invitation? It finally emerged: Robert May, the reigning power at Princeton ecology since MacArthur's death, fancied himself a considerable athlete, including in track. He annually challenged—and usually beat—staff and older graduate students in what was grandly termed "The Eno Mile." Guests of May's age who could beat him were avidly sought by the students, but thus far without success. I was their hope, for while a good ten years older than May, I was still, occasionally, able to manage a five-minute mile, faster than May's best. Unfortunately, I had a marathon coming up and was definitely not primed for short, fast efforts. Dan's pleas prevailed however, and on arrival, Martha and I went directly to the track to warm up. Her role was to be my "rabbit," to keep me on the proper pace for the first three laps of this four-lap event.

The race was formally conducted, with a clerk calling the roster and assigning us positions. The field was large, so this took considerable time. May, dancing up and down, grew impatient. "Just let the women start, we'll catch 'em anyway," he barked. My gentle, noncompetitive wife moved close to me. "Set your own pace, dear," she whispered. "I'm going to kill him!"

As the gun sounded, two fleet young students shot out of sight on their way to 4:30-ish miles, clearly in a different race. May led the rest of the pack, with Martha on his outside shoulder and I struggling to keep up several paces back. After each lap May would glance over at Martha, grimace, and quicken his tempo. She simply stayed glued beside him. After the third leg, I began to feel comfortable and surged past the two of them. Martha fell in behind me and we kicked around the last two turns, leaving May ever further behind us. When he finally reached the finish line behind us, he jogged on through it and off the track, disappearing for the day.

The students were so elated at the fact that a woman older than May had beaten him that the awards ceremony was delayed to allow for the purchase of a special Princeton mug, which was presented to Martha.

As for May, I never did get to speak with him. He showed up at my seminar but sat in the back row and ostentatiously read a newspaper while I spoke.

## THE SOCIAL CONSTRUCTION OF ETHOLOGY

The "objectivistic study of animal behavior" is how Lorenz defined ethology, but does the adjective "objectivistic" mislead? Can a science, especially a science with such obvious implications for society and politics as ethology, be immunized from subjectivistic infections? W. H. Whorf (1950) argued that our ability to form particular concepts was greatly influenced by the grammar of our language. Indo-European languages, with their requirement that sentences consist of nouns and verbs, actors and actions, are difficult to use to describe phenomena that do not readily dissolve into a subject performing an action, such as thunder or lightning. Similarly, some Amerindian tongues disallow the counting of nonpalpable entities—days or hours. In both cases, there is a tendency to adapt one's conceptions, if not the underlying perceptions as well, to the demands of one's language. Hardin (1956) has tracked the influence of the phrase "protoplasm metabolizes," which is a sentence akin to "lightning flashes," on the development of our current understanding of energetics in biology. His view is persuasive that the development of a more sophisticated view of energetics was retarded by this language. "Objectivity" also must suffer when research depends on funds from governmental granting agencies and publication or peer reviews. This is not to accuse the National Science Foundation or National Institutes of Health of anything nefarious. Indeed, as a sometime member of grant-review panels for these and other agencies, I am convinced peer review is as good a winnowing system as can be devised. Yet the fact that certain themes are "in" and others not inevitably influences decisions, as do personal beliefs and biases. I recall a "pink sheet," the critique by a reviewer, which commented upon a proposal of mine for underwater observations; on it was written: "P[rimary] I[nvestigator] cannot possibly supervise this research adequately as he cannot swim and is afraid of water." In fact, in my desire to do this work I overcame my fears, learned to swim, and became certified for SCUBA diving, but the critic was bound by his past memories of me. Then there was the occasion when the editor of a prestigious scientific magazine returned a manuscript I'd submitted (on why girls were particularly drawn to horses, in contrast to boys) saying that this was "an inopportune time" to publish on this topic. I presumed he meant politically inopportune, and he was correct, for it took some time to find a publisher, even though the substance of our article was never seriously faulted. How often does the anticipation of such problems influence one's choice of topic or the formulation of a question? Senator Proxmire's famous Golden Fleece awards, far from reducing wastefulness, have quite possibly contributed mightily to the formulation of grant proposals that are politically correct and intelligible to laymen. These are not always the most interesting options, but who dared buck Proxmire?

Anne Fausto-Sterling, among others, goes beyond these obvious influences and claims that *all* scientific research grows from social constructions that do not have their origin in operationalism or objectivity. The search for dimorphisms in the structures of the brains of males and females is one such example of an entire field of inquiry that grows from our society's conviction that, with few exceptions, people are one sex or the other. Fausto-Sterling actually cites data supporting a view I'd argued ten years earlier to the effect that sexuality is a multidimensional phenomenon.

Thus, what a culture means by male and female or homosexual likely may have nothing to do with the biological reality. Sex labels may not even refer to the same group of characteristics in different cultures. The biological reality may be that there are not merely two sexes, but several extremes with regard to sexual status, with individuals forming a continuum from one extreme to the other. Nor is this a simple continuum either, but a continuum along a number of different dimensions, so that what we are dealing with are *varieties* of sex. Each individual may be invested with a variety of degrees of maleness or femaleness. The factors that determine where an individual will fall on any one of the several continua representing the various kinds of sexuality may include the kinds of chemical conditions impinging on the developing embryo or the social conditions to which the infant is exposed. We have grown accustomed to a multiplicity of mating types in protozoans. We may yet learn to accommodate a similar view of our own kind. (Similar claims in areas more removed from the body politic than sex have been made by Latour and Woodgar [1985] and A. Pickering [1992], for immunology and particle physics, respectively.)

Basic to much of the confusion within ethology on how to define and treat such intangibles as "motivation," "instincts," and "innate releasing mechanisms" is a Western inclination, probably traceable to Descartes, toward reification. Whorf (1950) is doubtless correct when he sees the Cartesian view reinforced by Indo-European grammars. Could Descartes have been a Hopi or a Hindu? Hardly. In all events, the ethologist's customary solution has been to describe mechanical or electronic or mathematical models of, for example, the innate releasing mechanism, or "motivation," and then to search for the corresponding structure or "engram" within the central nervous system. Some of these models have a beguiling quality, not least because of their heuristic value. Consider the von Holst–Mittelstaedt "Re-afference Prinzip," which explains the perceived constancy of our surroundings. My characterization of this principle is to assume that you are standing quietly with your arm against the bough of an apple tree. In the first instance, you decide to shake the bough, whereupon a motor impulse (the *efference*) causes certain muscles to undergo periodic contraction and relaxation. Information as to the state of each muscle is relayed back to the brain from proprioceptors (the *afference*). In the second instance, while your are standing motionless, a breeze caught by the bough moves your arm. Assume that the identical muscles are moved and contracted to the same degree and in identical temporal sequence as in the first instance. Again, a proprioceptive input, or afference, is provided. In this latter instance, however, the ultimate source of the input was an agent external to yourself, the wind. The afference produced by the action of the external environment can be labeled *exafference*, and that resulting from voluntary movement *reafference*. The temporal and spatial pattern of proprioceptive impulses may be identical in both cases. Nonetheless, our perceptual apparatus distinguishes between voluntarily moving your arms or movement by an external source. Thus, the central nervous system imposes a meaning on the reafference different from that imposed upon the exafference. This, in turn, is evidence of central activity that proceeds independently of the periphery.

Von Holst's model assumes that every efference leaves an "image" of itself at some lower center to which the reafference must compare as a negative to a positive print. Superimposition of the negative and positive may erase the image, or *efference copy*, but should erasure not occur, predictable illusions may follow. Thus, if your eye is mechanically fixed with the muscle proprioceptors narcotized and you command it to turn to the right (i.e., form an efference copy), the visual field will appear to jump to the right. No reafference could result, of course. Now, let your eye be turned to the right mechanically. This time the field appears to move to the left. Movement of images across your retina produced a simulated reafference but there was no efference or efference copy. If these two operations are simultaneously executed, i.e., your eyes turned right mechanically just as you command them to do, the visual field appears stationary.

This model has proven useful in the design of experiments that have further clarified the nature of the processes that lead to perceptual constancy. Inevitably, however, it has also lead to efforts to find the physical correlates, the substrate, upon which the process is built. The model, the construction, is now shaping the research, and thus the results.

What is "objective" reality in ethology, anyway? Can any behavior be examined in the absence of a priori social and linguistic suppositions as to its structure, function, and mechanism? Mathematician Anatole Holt used to say that the job of the scientists was to view the world through many different kinds of glasses. Each lens, he said, gave one a slightly different perspective. It is the development of a constellation of different views that he saw as the business of science (Bateson 1972).

Even within our Western traditions, "science" is not an unchanging monolith. Yet alternative or complementary schools rarely coexist for long. Ethology had its start in the thirties, but other schools dedicated to behavioral studies vied with it. Pre-Lorenzian behavior studies were generally characterized as dichotomized: on the one side, the vitalists, for whom unanalyzable "instincts" were the components of complex behavior; on the other, the behaviorists with their reflex chains. The attractiveness of Lorenz's theories, it was claimed, lay in their integrating ideas on motivation and control with behavioral processes drawing on the neurophysiological work of E. von Holst, and particularly his demonstration of spontaneous activity in the central nervous system. However, this is a grossly incomplete view of the behavioral scene. Biologists and psychologists such as Carl Lashley, Frank Beach, and Ted Schneirla in the United States and William Thorpe in the United Kingdom had already broken new ground that lay elsewhere than along the paths of the two traditional schools.

Lorenz's success, I believe, did not lie in his supplying the only reasonable alternative to existing views. Rather, by formulating a new vocabulary that identified the important elements on which his theory drew ("vacuum activity," "releasor," "action-specific energy," etc.) he facilitated discussion at the same time that he forced descriptions into terms that suited his preconceptions. Moreover, and likely of greatest importance, as countless of his students and admirers attest, his vivid characterizations of examples derived from his own experience, sometimes acted out in front of a podium, other times compellingly told in his many delightful books, tended to dissolve doubt or criticism. "Charisma" is not nearly a strong enough word for his persuasiveness!* Further, critics were often put on the defensive or silenced altogether by Lorenz's rages against those "nincompoops"

---

* Another, rather different example comes to mind. Konrad had an enormous saltwater aquarium in his bedroom, complete with a living portion of coral reef and associated creatures. One of his "pranks" was to select a young and timid woman from among his guests and, ostensibly spying a loose piece of seaweed, strip off his clothes and give them to his guest to hold. He would jump into the tank and, paddling about in the nude, would seem to make repairs. He would then emerge and dress as if this were a daily occurrence. So far as I know, no one ever said anything while the act was in progress—only later. Given his ability to distract and disarm, Lorenz would have made a great stage magician.

who couldn't distinguish a turtledove from a pigeon but would not be satisfied with anything less than a quantitative description of their behavior. Finally, Lorenz's extension of his views to humans and his ability to articulate them in language accessible to the laity certainly helped to identify him as a potentially important integrating figure. Thus, for several decades, the Lorenzian approach has dominated much of behavioral biology.

Latour and Woodgar (1986), in a detailed study of the day-to-day operations of a biology laboratory, forcefully make the case that it is indeed the personal attributes of the investigators that are the primary factor, as compared with formal theories, in the development of scientific "facts" or concepts. "Informal communication," they write, "*is the rule*. Formal communication is the exception as an *a posteriori* rationalization of the real process."

They continue, "most published papers are not read, the few that are read are worth little, and the remaining 1 or 2% are transformed and misrepresented by those who use them. . . . Each text, lab author and discipline strives to establish a world in which its own interpretation is made more likely by virtue of the increasing number of people from whom it extracts compliance. In other words, interpretations do not so much as *in*form as *per*form.".

Hacking (in Pickering 1992) carries the argument one step further: "[As a] science matures, it develops a body of types of theory and types of approach, and types of analysis that are mutually adjusted to each other. They become . . . a closed system that is[,] essentially, irrefutable. . . . High[-]level theories are not 'true' at all."

What, then, do such reflections portend for the future of ethology?

## THE FUTURE OF ETHOLOGY

My musings on the construction of ethology attribute the dominance of particular ethological theories to the persuasiveness of particular protagonists, specifically, Konrad Lorenz and Niko Tinbergen. There is evidently nothing inherently more "objectivistic," "empirical," or "heuristic" about their ethological views than those of many others, nor can it be freed from the social matrix within which it is embedded. But as this could be true of Western science altogether, why single out ethology for this criticism? Surely, so long as a scientific enterprise is productive within a given culture, its less than universal character might not be of concern.

Ethology, however, has achieved a public status that does dictate special concern. On the one hand, its subjects are organisms that are perceived as representatives of an evolutionary lineage that culminates in human-kind. Consequently, generalizations on the causes and consequences of animal behavior, it is believed, *must* have implications for theories about human behavior. On the other hand, unlike psychologists who explicitly study human behavior, the proper subject for studies by ethologists is not culture-bound *Homo sapiens* but the relatively uncontaminated (by human culture) "lesser beasts." Ethological pronouncements on the genetic deter-mination of, say, aggressive behavior have a cachet that pronouncements by psychologists (who study people) lack—as witness the ready attention given the sociobiological theses advanced by Lumsden and Wilson (1984). Hence, I believe it behooves us to treat ethological research and pronounce-ments with a particularly high measure of skepticism: the consequences of faulty interpretations are neither as readily apparent or as easily corrected as in particle physics or genetics (although environmentalists worried about the release of genetic material to inappropriate hosts might disagree).

In 1989, a group of eminent ethologists gave their individual views on the likely future path of ethology (viz., Bateson and Klopfer 1989). Among them was George Barlow, a one-time student of Lorenz and certainly one of the most articulate, productive, and influential of the sixties generation of ethologists. His chapter title, "Has Sociobiology Killed Ethology or Revitalized It?," takes its theme from E. O. Wilson's *Sociobiology*. With the advent of sociobiology, "animal behavior was to fall through a yawn-ing crack between sociobiology and population biology on the one side, and neurobiology on the other." Barlow, however, finds that sociobiology, by raising new questions and refining old ones, has served to revitalize ethology. There is a distinction in the approaches of sociobiologists and ethologists, which Barlow sees as a barrier to either the assimilation or displacement of ethology: the former, as Kamil (1983) put it, study pro-cesses such as adaptation and optimization, which help in "understanding" or predicting behavior, but for ethology it is the behavior itself that is the object of study. Even some of the classical behavioral studies, Barlow sug-gests, quite apart from new ones indicated by sociobiology, warrant reex-amination with a more sophisticated contemporary methodology. The new ethology, in short, would appear to Barlow to be a technologically im-proved variant of the old.

Marian S. Dawkins (also in Bateson and Klopfer 1989), who also has thought about ethology's future course, does not disagree with much of Barlow's evaluation but argues for the need for more studies of mecha-nism. The four legs of ethology—mechanism, function, development, and phylogeny—have been unevenly developed. "With a bit of readjustment of its weight onto all four of its legs," she asserts, ethology can continue to serve in the future.

Another prognosticator, John Gittleman, sees ethology's greatest weakness, and the key to its future, in the inadequacy of its comparative database, particularly in a bias towards particular taxa (birds) or particular behavior patterns (nesting patterns). He also emphasizes a need for "an abundance of standardized, quantitative, and long-term single-species studies."

It seems to me these three views are fairly representative of how the community of ethologists views their future. Unspoken, but I suspect equally a part of the consensus, is the abandonment of the Lorenz-Tinbergenian schemata and the absence of specific replacements. The Great Men are dead and no one seems likely to claim or be elected to their thrones. Perhaps ethologists have become too numerous and diverse for its future Nobelists to have as tight control over the field as did their forebears.

Also unspoken is the implicit agreement that our present constructions of the phenomena we study require little review or revision. If our way of viewing the world does not assure glimpses of the ultimate realities, it is at least the best of all possible views. To be sure, there are critics: I think of Linda Fedigan or Anne Fausto-Sterling, but to object (as do they) that we arbitrarily construct our subjects-/objects-of-study is a long distance from suggesting how we might otherwise enhance our understanding of the world around us. The unresolved questions about animal orientation, navigation, cognition, and the like might yet be answered as our analytic means improve. Then, again, these improved methods may also fail, reinforcing the argument for new paradigms, altogether new ways of putting our questions. How do we develop these paradigms? Is there a strategy that can be developed or must the enterprise be left to individual intuitions? This issue, I think, is the main challenge facing ethology, and, indeed, many other domains of biology.

A more specific dilemma we in ethology must resolve has to do with the subject of our study: behavior. It has become accepted practice by all but the most fastidious editors to allow what had once been a collective, or plural noun, to stand for the singular: a behavior. But what is "a behavior?" Surely our interests extend beyond the external manifestations of certain muscle contractions? But even if they did not, few actions are so stereotyped and invariant as to be cleanly separable and individually distinguishable. I suppose one can count eye blinks and record their frequency and duration. "An eye-blink behavior" could then be constructed. Or, one might measure pupillary dilation in response to the presentation of different images, noting response latencies, magnitudes, and duration. The construction of "a dilation" becomes greatly complicated by the addition of the third parameter. What is required before two successive responses are considered different?

More complex acts, as in courtship or territorial displays, the patterns that are of greatest interest to ethologists, simply defy any but the most crass efforts to be disarticulated as one might disarticulate the bones of a skeleton. There is a certain stability to the overall pattern, but its components lack the structural fixity that exists between a whole skeleton and its parts.

We are coming close to the one-hundredth anniversary of the idea that behavior can be homologized in the manner of anatomists dealing with bones, and are already past the fiftieth anniversary of Lorenz's bold claim that this was the Archimedean principle of ethology, without being a whit closer to being able to profit from the principle of homology.

The nature of "behavior" must be reconsidered; fixed action patterns are a will-o'-the-wisp, and mere reflexes, with all their neural complexities, provide insufficient grist for the ethological mill. We need to start anew to define and delimit, in a meaningful way, the subject of our discourse, behavior.

Gregory Bateson once remarked to me that he was occasionally dismayed by the thought that the universe was so much more complex than the human brain that the latter could never hope to even describe it: instead of having seven blind men describing an elephant, which at least allowed for the possibility of a synthetic description, the brain was as a single one-armed and legless observer. Not an encouraging metaphor, but as Bateson never did cease thinking about and studying the world, it seems clear that he was not discouraged by it.

## PROBLEMS OF THE YOUNG SCIENTIST

A number of my recent (and much younger) students have occasionally asked whether I would make the same career choices now that I had made (not necessarily with deliberation) in my past. Job competition now is fierce, tenure is often an elusive goal, and teaching loads continue to rise.

In my final year as a graduate student I received a letter from the president of a well-known Eastern university, offering me a post in his sociology department. I responded with a polite note explaining that I was not a sociologist, but a zoologist, and did not yet have my thesis completed. A telegram came in response offering an assistant professorship in biology. This was not what I'd bargained for when I commenced my studies, however. Academic posts were scarce and poorly paid. My younger mentors at Yale had to seek summer employment in local factories to sustain themselves. My expectations were for a poorly paid post at a preparatory school, or, worst case, a public high school.

The post-Sputnik era changed that. By the time I'd completed my doctoral work, funds for postdoctoral fellowships had begun to flow; universities—in response to a rising population of eighteen-year-olds as well as federal tuition subsidies—expanded and new openings seemed to exceed the number of qualified applicants. Young academic aspirants could pretty much go where they wished, and teach and study what they wanted. Salaries were low, but adequate and assured, with ample fringe benefits. Tuition remission for faculty children was common, tenure taken for granted.

The outlook today is rather different: few jobs—the consequence of retrenchment and the abolition of compulsory retirement— limited security, and diminishing fringe benefits, to say nothing of the paucity of research and travel funds for most young scholars.

But, as I remind my questioners, that too is what I had to look forward to. It turned out differently and I've savored that, no doubt about it. But I've no doubt either that my path was set before I had any indication of the largesse to come.

There were other problems I did not escape, problems not due to the political and economic circumstances of my time, which continue to afflict young scientists today. Lawrence Kubie, a psychiatrist of note, discussed these in a pair of articles published in 1953 and 1954 gathered under the title "Some Unsolved Problems of the Scientific Career." Forty years later, I'm struck by how apropos Kubie's diagnosis remains.

Kubie's "problems arising from socioeconomic forces" appear to me to be as real today as when Kubie wrote. Three issues were especially identified: economic privation, divorce from reality, and the uncertainty of success. The first of these is easy enough to identify, and hard to correct. Even if salaries for young academics are generous, these become available only after a long period of struggle: four to seven years of graduate school followed by several postdoctoral fellowships, during which time debts inevitably mount. Where parental resources mitigate the financial burden, a continued dependent relationship may inhibit the development of maturity. Unlike the outcome of training for other professionals, once the goal of a tenure-track position is achieved, a rapid rise up the career and salary ladder is still not likely.

The long, intense educational period poses other perils. Opportunities for engaging in normal social and political activities are necessarily limited; the longer the educational period, the more intense the selection for a cadre of scholars who are content to ignore the larger world and the more homogeneous and science-centered the pool of peers becomes; and, because teaching is usually a part of graduate training, opportunities for lording it over younger, naive students abound. As Kubie notes, teaching science to novitiates is heady wine and cannot fail to stoke vanity. None of this is conducive to the development of maturity.

Finally, there are the special strains generated by the uncertain outcomes of scientific experiments. So much can go wrong. And even when nothing does, someone else may publish first. The second discovery of calculus, however independent, brings no kudos.

The point here is not so much to dwell on the impediments and "wind shears" that Kubie identifies and whose effects he analyzes, as to suggest that little has changed: then, as now, the cost of a career in science is high and the external rewards are uncertain.

Yet, if I am honest, I would admit that I never "chose" a career in science, let alone ethology; it was more a case of having been chosen. My divinity school friends speak of the ministry as a vocation, a "calling" to which one is compelled to respond on pain of eternal dissatisfaction.

As to those students' queries, my answer, I think, is evident.

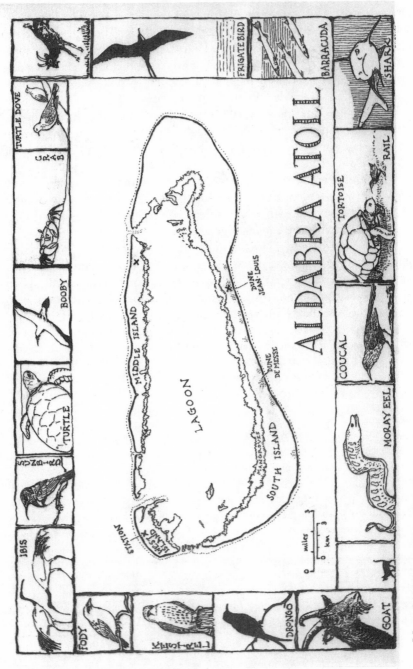

**Map of Aldabra Atoll by Kate Davis**

# Aldabra Journal

*10 December. Mahé (Seychelles)*

Lunch over, we drove to the pier where we found our five-hundred-ton tub, *The Nordvaer*. No sooner aboard, however, we were asked to report to the captain. Sorry, he explained, first mate is ill and must go to hospital. Under Seychelles rules, no freighter lacking its mate may carry passengers. But, since we were the only passengers, an "out" was available. We could sign papers and join the crew. We could? We would! We did.

Regrettably, we were not made first or even second mates. Instead, we were listed as "supers." That implies neither praise for our characters and ability, nor does it stand for superintendent. It meant, we learned, "supernumerary." We'll not stand watches or handle the wheel, sad to say. But, we can at least ignore signs that say "no admission, crew only."

The crew's favorite occupation appears to be drinking and fishing, and the rusted, unkempt condition of our craft attests to both sports. For a time this a.m. bonitas were being hooked as fast as the lines were being thrown out, but now the school has passed. We saw none of the fish at dinner. Fare is English at its worst, tinned stew, old fish and chips, a disappointment as we'd been led to expect more Creole cooking. Little likelihood we'll ever overeat. The water-weak coffee (another shipboard myth shattered) will also help wean us from coffee for a time. But we're still enjoying the lazy, loafing shipboard routine, interspersed with some serious seascape study—lots of albatross, fish, and ever distracting cloud sculptures.

Flying fish schools this morning. We first thought them to be birds flying low—unbelievable distances they cover in the air. We've timed them up to ten seconds and at a speed not inferior to ours. We could see them change the angles of their fins as they "flew"—whether this actually was of functional benefit in prolonging their glide, we don't know.

*13 December Monday (we think), Indian Ocean*

The shipboard lethargy grows—even arising for breakfast takes an effort, though we've no shortage of sleep—and even slept comfortably, thanks to the air conditioning. Neglected to comment yesterday on the dearth of atolls, and the reduced signs of wildlife, except for the flying fish. Nothing was caught by our stern-side fishers, and birds are few too—only an occasional booby and gannet. Aldabra just came to view, one day earlier than expected. No one here knew quite how. We suspect it was the alcohol that was supplementing the diesel fuel, and *sans* first officer the ship's company may have improved its efficiency.

### 14 December

We're here—settled in a small, airy cell next door to Meg.* Conditions are simple, clean, primitive, but adequate. There are decent scientific quarters, which are even air conditioned. Eighteen scientists and technicians on board, plus Seychellois laborers, but half are at the two field camps on the far side of the atoll.

Birds are incredibly tame—of the three dozen species, we've already seen one-third—including a kestrel nesting by our door, coucals, drongos, fodys, and sunbirds. To take photos one adjusts the desired focus, counts to twelve, and waits for the birds to show.

The main campsite problem is water shortage. The solar still is not functioning and rain catchments are empty—rains have fallen all around but not on the station. So we're being rationed—all bathing, etc., in the sea (we were issued saltwater soap) and heavy on beer, easy on fresh water at table. The problem is deadly serious for there's absolutely no alternative water source. Ironically, we can see the rainfall on the ocean to the south.

After unpacking, we jogged a mile or so down the beach. There is a ten- to twenty-meter swathe of white sand along the west end of West Island, when the tide is out—hard going in the soft sand but we persevered to the northwest corner where the beach gives way to a two-meter-high coral cliff—undercut by waves at midtide and thus full of caves (and crabs). From sand beach seaward was a shallow lagoon—one-half to two-thirds of a meter deep, crystal clear, perhaps one-half km. or so to a reef over which the waves were breaking. We shed our clothes and sported about like a pair of porpoises, then just sat in the warm (30ish C) water to our chins, enjoying the white sand, black coral cliffs and all around us tiny fish, transparent or brightly hued. We kept a sharp lookout for moray eels, which abound, but despite some likely holes saw only imagined ones. A frigate attacked one of the boobies (there are three) directly over our heads and a group of pied crows chorused from behind the coral. Among the rarer birds on this planet is the Aldabran white-throated (flightless) rail. One of them lives directly on the station grounds and is inclined to peck at exposed toes.

Remarkable how one gets the same feeling from the exotic vegetation types here as from their more familiar counterparts at home, even though the plants are totally unrelated. The casuarina here is an extremely primitive plant, botanically related to the horsetails *(Equisetum)* and to walk through a grove of them is just like being in a pine forest at home except perhaps for the lack of scent. The ground

---

* Meg Gould, Duke zoology graduate student.

**Aldabra Ho**

**Main Station**

**The Pemphis (with Meg and Martha)**

is covered with needles and small cones. The wind sounds the same in tne jointed needles as in the ordinary ones. Also, the scrub growth over most of the island has much the same quality as the Western chaparral but no manzanita and sagebrush here, just lots of other stuff we've never heard of. Much of it is sharp and thorny.

### 15 December

End of first twenty-four hours in the field. We've great respect for Meg. Her workday begins just after five a.m., ends after eight p.m., and involves the roughest terrain and hottest climes imaginable. But now we're relaxing while awaiting the teapot's boil, watching several dozen sharks cruising about the lagoon ten meters away. Swimming?

Yesterday we rose at six for a short run down the beach, a brief dunk in the ocean, breakfast, and then took our gear in a large dory for the two-and-one-half-hour ride to Middle Island. Except for the push through the reefside surf, the trip was smooth, enlivened by schools of porpoises with frigates and boobies accompanying overhead. We cast fishing lines astern, and, after half a dozen barracuda, began hooking all kinds of bright-colored beauties. "C'est bon," said one of our Creole boatman, smacking his lips, and "C'est très bon" for one of our more colorful catches. Two tuna were landed, but our prize was a forty- to fifty-kg. (maybe more) kingfish, over one meter long (have photo to prove it—also a bandage where the fish line cut my hand). A storm enveloped us as we beached, our first rain and badly needed, but in a few hours time all was clear and a stew pot bubbled away with a fine Creole fish stew. At three, Meg and we left the others and began a goat-stalk that continued until dark. Supper after eight and an exhausted, too brief, sleep.

Field camp consists of a three-sided corrugated tin shed, two by four meters, where the food and gear are stored and whose roof provides a water catchment. Tents on the small sand beach in front of the shed, surrounded by coral and mangrove, are where we sleep.

From ocean to lagoon, there are four distinct zones into which the eight hundred meters of the island's width are divided. The shore at this end of Middle Island (it differs greatly from one area to another) is marked by casuarinas and resembles a temperate pine grove, needles on the floor, clear trails and cool breezes. Crabs scurry everywhere and grunting tortoises are wherever one cares to look. Early and late the trees are festooned with awkward-looking frigates, like oversized Christmas-tree ornaments. Twenty meters inward, the "Platen" begins, flat, weathered coral, fairly open, one- to three-meters-high vegetation and moderately easy walking. But then come the Pemphis, three to four meters high, so dense as to be totally impenetrable without machete (sorry, *pongo* is the term here) and with no visibility of goats more than one animal-length distant. One trail has been cut through this stuff and another's abuilding. The mangrove layer is the last—only a few meters wide, a tangle of roots arched by broad-leafed crowns—then the lagoon with its mushroom coral columns (the champignons).

The tameness, or rather fearlessness, of the animals is, of course, proverbial on uninhabited islands, but still amazes us. Rails, common pigeons, turtledoves, and on and on. Hardly any fun to photograph them. Viewers will assume that a telephoto was used or that they were hand-reared.

### 17 December 10:30

After a five-hour goat watch, we returned for a cup of tea (always). Midday is a good time to be thinking over projects while sitting in the shade. The research project we will now describe for you has to do with tortoise tickling.

Yesterday, Meg showed how the huge land turtles would respond by rising up on all four legs when tickled inside a hind leg. There's a great controversy as to whether this is a sexual response or whether it is done to aid the little flightless rails that catch flies around tortoises and may use their long bills to clean parasites from the animals' skin. Some observers have seen ectoparasites on the tortoises and at least two others have seen tortoises rise up at the approach of a rail. Another theory is that the tortoises are getting themselves higher to protect their eyes from a possible peck, but it seems it might be simpler just to withdraw the head into the shell. Our idea was to find out whether both male and female show the response, reasoning that if only males do it, it is probably a sexual response. So Martha spent a couple of hours trying to quietly stalk tortoises from behind and give them a tickle. "Science" comes in strange forms sometimes. Problem #1 was not alarming

**Martha testing the tickling response**

the beasts, for then they'd draw in the hind legs quite suddenly, hiss, and move off. (M. employed a stick as a tickler after getting her finger rasped with a claw or shell edge a time or two.) Problem #2 was to find out the sex of the beasts. Males have a larger and fatter tail and a more concave lower shell, to facilitate balancing on the female back, presumably. But, both of these traits are relative and are less clear in the smaller (younger) animals. The larger ones were too heavy for examination most of the time. So, M. did a lot of guessing and tried to find beasties with I.D. numbers from a census someone made. We can hook up these numbers when we get back to the station and sex them then. M. got responses from a few of those whose sex she was pretty certain of—it does seem to be a trait shared by both sexes.

While the coral is no less hazardous for old-timers, one does become more daring and casual with time. Familiarity breeds contempt. Just how contemptuous we have become may be gleaned from the fact that we caught goats last night (for marking with ear-tags), which is done by cornering them on coral ledges that overhang the ocean (also sharks!) twenty or so feet below on jagged, razor-sharp coral with narrow pinnacles separated by deep, foot-grabbing crevices, and in the dark. We did survive and even tagged three goats and made measurements on an additional three.

### 19 December

The long delayed rainy season has arrived. We had a short shower, lots of wind. During the night we thought we might emulate Dorothy's trip to the Land of Oz, but luckily our tent was well moored and the canvas strong. Lots of lightning activity too, scary on an open island.

Today's goat walk has us along the beach at high tide so we can see the blowholes in action. Rather dramatic. They are often set some distance back from the cliff edge, twenty meters or so and the great roars or hisses and towers of spray or mist they emit, often three to four meters high, seem unrelated to the sea's movements. Better than Yellowstone!

### 20 December early p.m.

We're sitting in the shade of the thatched hut at Dune Jean Louis trying not to be a source of disturbance for the sunbird whose nest hangs from a piece of thatching material about a meter from our heads. Dune Jean Louis Camp is reached from Middle Camp by puttering across the lagoon in one of the outboard boats. One can hike around the eastern end of Aldabra and get there on foot, but it takes a couple of days. The boat rides take an hour if one calculates the tide correctly and there's enough water and if the motor on the boat doesn't malfunction. Everyone has tales of times when they had to hoist a shirt or an oar by which to sail, or row. Beaching is scary, wading in waters too murky to keep a lookout for sharks, and pushing the heavy boat.

Dune Jean Louis is indeed a big sand dune. It's the highest point at Aldabra (sixty meters).

**Meg in our hut at Dune St. J**

*Late p.m.*

Harry (our Seychelles assistant) has just returned from a dinner expedition. One has to bash some burgher crabs, then pulverize them, mix well with sand (saving one or two larger pieces); the mixture is cast into the surf (which is reached by wading though shallows swarming with moray eels—great sport). The baited hooks (with the large morsels) follow. After the first shark lunched one-half meter from Harry, P. gave up and retreated to the rocks. When, minutes later, the shark returned and cut Harry's line, he followed. So we chased a school of tide pool fish, twenty to thirty cm. long, into a corner, built a coral dam of loose bits by hand and had a fish bash, hand to head (*pongos* actually, dull side). Resourceful, these islanders.

*21 December*

Goat watching at Dune Jean Louis is of the stationary sort, from a height. What with Harry's two good meals to work off, M. and I jogged, walked, stumbled and crawled along the coast, some sand, some low grass, lots of jagged coral. On our return, tide was out so we cut across the exposed flats. After several brushes with fair-sized morays, we returned to the slower tour on the heights.

Hundreds of crabs. Plovers also in evidence this a.m., along with other migrants we couldn't identify—no glasses and unfamiliar species makes a poor combination.

**Our subjects. Photo by Meg Gould.**

Also saw a cat, thin, black. Really sorry we couldn't shoot it—there are only a few left on the islands and, fortunately, no more dogs, but the cats can devastate the flightless and ground-nesting birds. Norway rats are another imported pest, but these are so incredibly numerous that their eradication is out of the question. Presumably they've been here long enough to have attained some sort of equilibrium with other species. We saw some recent green turtle nests, but as yet no hatchlings. They are apparently produced all year round, to the evident delight of the scavenging pied crows. Once lunch has settled, we'll head north (towards the lagoon) seeking our capricious friends—none were sighted in either last night's or this a.m. watch, despite ample fecal evidence of their propinquity.

*22 December*
Still no goats. We hiked west after our watch, and found lots of feces—some fairly fresh—so tonight we'll spread down the coast. Lots else to watch. Last night we came upon a mammoth green turtle digging her egg pit in the sand, and this a.m. a second fresh pit was in evidence, too. Meg found some green hatchlings emerging and rescued them from the pied crows that were awaiting their arrival. (Probably the sharks ate them instead—few young survive.) This p.m. Meg saw two greens copulating, then saw the male stranded, the price of being *amoureux* on

a receding tide. With Meg steering and me hoisting the hind end of the two-hun-dred-kg. beast, we wheelbarrowed him into a tide pool, then covered him with seaweed for insulation until rising waters should bring him release. I guess I can now have a cup or two of green turtle soup with a clear conscience. Turtles are a forbidden food here.

Also newly seen, a just-hatched giant tortoise all of four cm. long. It looks ridiculous alongside the one-and-one-half-meter-long adults that litter our camp-ground. (Yes, you can ride them!)

As to the latest in gourmet specialties, Aldabra style, today's was by far the best of a series of delights, "Satinee" Recipe.

Remove shoes; enter surf to knee depth (not more); kick vigorously, keeping sharp watch. The highly aggressive white-tipped sharks will soon appear. As they close in, bash one or two with a heavy knife (a *pongo* is great), grab them by the tail and flee. Filet the shark (two, if you're serving a crowd), cut into four- to six-cm. chunks, wash several times, boil till meat falls apart, wash, press dry, then fry lightly with all manner of freshly pounded spices (pepper, caraway, vinegar, curry, etc.). Marvelous with rice! Shark is now a regular item on our bill of fare. (The meat is light in color, fine in texture, and mild as the best white tuna, once the urea is washed out.)

### 23 December

Another disappointing pair of watches. Last evening we flung ourselves down the beach, one of us each on the three major dunes. Great visibility but naught in the way of goats to be seen. I arose about five this morning, well before morning light, and stumbled back to my dune outpost to catch a glimpse of early risers. An early lightning-rainstorm reduced the visibility for a while, but by 6:30 it cleared and lightened enough that I could confirm the absence of goats. Meg thought she had spotted a single beast far in the distance, but that's of little help. On his return, I hiked across the lagoon, but no sign in the Pemphis either.

### 24 December

Back to station, following an uneventful two-hour crossing. Did see several rays—and one green turtle swimming in the lagoon. The sights and sounds (popu-lar music tapes, ugh) of civilization are not attractive after the near solitude of our camp. We find ourselves resenting the presence of other people. Great Christmas spirit, eh? Perhaps the plastic tree and tinsel in the dining porch will displace the hermit urge.

### 25 December. Merry Christmas

Ours has certainly been unusual. We've found the return to civilization, i.e. base camp, somewhat jarring—loud, unremitting "music" (damn tapes!), much alcohol, and silly talk from the meteorologists ("met-men") and technicians who dominate life here, since the scientists for the most part spend their time in the field. Tinned food was also a comedown after Harry's marvelous seafood treats, ough the fresh bread was welcome. We escaped briefly with a run down the ile-long beach on the west coast and a swim in the seaside lagoon—here largely ark-free, thanks to a clear, unproductive sand bottom and a reef one hundred eters seaward. Christmas Eve followed, celebrated in what we were reminded as traditional English style: a drunken bash. It was held (as a mark of courtesy) the home of the Seychellois headman—much dancing, animated but drunken. . performed heroically, for there is a great shortage of females—of the one-and-ne-half dozen Seychellois, only three or four are females and the "Europeans" that includes us) only add six. We sampled the local home brew, *kulu*, which we'd bserved in the making earlier. It's derived by fermentation from the sap of coco-ut palm collected from the base of new fronds high up in the trees. The only thing be said is that it smells far worse than it tastes, which is bad enough.

We stayed with the festivities, in order to join the drunken procession that ampled into the small, attractively decorated chapel at twelve, for R.C. "serice," improvised, monotonous, noisy, but still providing a fascinating glimpse to the lives of people trapped between two different worlds and centuries. The d mourn the passing traditions of English colonialism. The young, we've learned rough fragmented discussions, are developing rather considerable hostility to-ard the remnants of colonial rule the Royal Society represents, for all their smiles nd "Bon jour Massa, Madam." As to the management of the workers, their con-acts specify a fixed food ration and wages—no family supplements. A man with is family here must buy additional food and pay rent—no hope for ever saving ny money, for food prices are high, fishing is restricted, hunting is forbidden, and ages are low. In the Royal's defense, let it be added that conditions in Mahe are o better—jobs are scarce. On Aldabra, however, the workers' conditions contrast ather starkly with that of the "Europeans" who are but a small part of Mahe's life. he logic of our killing animals for study while forbidding taking them for food lso escapes the workers (me too, for that matter).

Up early in the a.m. for a solitary run and swim. We hiked down the beach gain, watching the butchering of the green turtle the Seychellois had been al-owed to catch for Christmas, and the removal of its eggs. Then, lazed in the sun in ne of our most private beaches, overhanging coral ledge bordered by coral cliffs hat one gets around only at low tide. Loverly. Once in, you stay until the tide goes ut, then you can wade around the walls. Don't go in with a rising tide, though. Ve've memorized the tide tables, you can be sure. With a tide of up to four meters, ne doesn't fool around; the tidal current can attain speeds to six knots.

### 28 December Dune Jean Louis

It would be overly dramatic, perhaps, to compare our lot with Robinson rusoe's, but for a brief time and relative to our survival skills, the analogy is lose. The main difference, of course, is we chose to be marooned in order to try to nravel the mystery of the whereabouts of the supposedly overabundant goats.

On the way over, we rescued a waterlogged tortoise, five meters from the earest shore. So, they can swim, but this one didn't look like he'd intended it.

*29 December. A long, weary day!*

Yesterday ended later than planned when we discovered ants nesting in th
timbers of our canvas shelters. With only one mosquito net to cover both of us, th
was the final straw, so we raised a tent by moonlight, improvised a padding on i
floor for me, as there was only a single cot for M.

Up early and westwards to Dune de Messe. Though only five km. distant, ou
steady walk took two and three-fourths hours; which should tell you somewh
about the terrain; different from Malabar, too.

The return trip took longer, both because we finally spotted goats and becaus
we more often forsook the spotted goats for a slower plod on the soft sand. Th
prevailing winds bent the grass spikes toward the west, so our return had us impa
ing tired feet and shins. The return trip was more pleasant in that it was gettin
cooler, not hotter. As we grew more fatigued, the southwest breeze brought us rai
from a nearby storm while the sun warmed our bare backsides (we are tan enoug
now to dispense with clothes almost all day, evening excepted—millions of pe
son-eating mosquitoes).

Unlike the west coast beaches, and the very few small ones to the north, th
southern beaches have much debris, carried there by southeast trades, mostly ship'
timber, including what appeared to be masts and yards of ancient sailing-sh
material. It decomposes slowly here (last month's goat feces still look fresh) thoug
animal remains are quickly dismembered and eaten. Next most common are fishn
floats (mostly plastic, alas; glass is rare) and flip-flop sandals every km. Not muc
else, actually—some well congealed oil blobs every now and again, of course.

Just now we're dressing wounds, coral-cut feet, and ugly burns of "blister
beetles—a flying cockroach-shaped beast that is attracted by smoke and releases
highly caustic, inflammatory (and then infectious) substance when squashed c
smitten. Later, perhaps, we'll jog around our clearing or jump rope. Dune crossin
is too strenuous now, besides which it accelerates dune erosion. As you may d
vine, we are slowly succumbing to tropical lassitude. Fortunately there is no ex
cess of either solid dainties or alcohol, so we hope to avoid the otherwise inev
table tropical pot. (Indeed, it's noteworthy that residents at base camp are phys
cally differentiated from those scientists whose work keeps them in the field.) W
do have one bottle of rum along, for medicinal purposes, of course, and will use
this evening as an excuse for some kind of rum punch. Happy New Year.

*1 January*

As anticipated, our New Year's Eve was indeed quiet, not even the surf cor
tributing, low tide and no wind. We finally fled the hut—too buggy and windless—
and sipped our potions on makeshift stools outside our tent. One advantage of suc
celebration is that there's no morning-after cleanup required. Not even our stoo
required dusting or rearranging (they were a pair of giant tortoises, peacefull
sleeping, as they do when it's hot or dark and they are conveniently disposed).

We've passed a day dozing, reading, and timing the incubation intervals c
our resident sunbird—now with two eggs. The male, who helped with nest cor

truction, now no longer enters the hut, though he accompanies the female on flights around camp when she leaves the nest (which is often) and sings from atop our nearby tent or clothes-rack bush. This may not be typical, however, or perhaps he's inexperienced. Her nest, a hanging globular affair, was badly attached and had it not been secured with twine, would surely have fallen.

### 2 January

We lay abed an extra thirty minutes this a.m., not reaching our watch station until six. Even then it was darker than night, with heavy black clouds blotting the lagoon. It did lighten, and finally, we did see some goats, distantly. As we set out on our fishing expedition we were submerged in a long overdue deluge. Wet and cold, we returned to the hut for shoes and exploited the cooler temperatures for a run. M.'s off on an eight-hundred-meter circuit on the short, tortoise-cropped turf, which winds its way around the coral fragments. The tortoises are a good height for me to practice hurdler's strides, though their mobility is somewhat of a disadvantage here. Shower and rainstorm over, we used some of the older water, now too foul to drink, for a luxurious sponge bath. "Clean" is relative, so by Aldabran standards, we're squeaky clean—even if still sandy.

### 3 January

Our last full day alone has dawned darkly, and thirty minutes after stationing ourselves in our watchtower three meters up, the lightning frightened us down. Atop the dune, we lasted longer, but saw no more goats.

Ibises, incidentally, at least the sacred sort, like the taste of cooked oatmeal. Turnstones don't. Tortoises like anything, from margarine to fish soup.

One of M.'s chief occupations is checking tides. The station hasn't received the current tide tables from East Africa yet, and, to our surprise, they are not so regular as to allow a simple extrapolation from those previously received. Since tidal fluctuations are over four meters, with currents of up to six knots, and many areas of the lagoons are totally dry at low tide, you can see their practical importance. On many days it would not be possible to cross the lagoon on one tide; with no lights or markers, nighttime travel is precluded. Thus, for all but two days of the year (uh, which days?) cross-lagoon trips take two days, with departure having to be timed rather accurately. A fifteen-minute error may then require wading for one to two km. in murky, knee-deep, shark-infested, and coral-sharp inlets. Or, equally bad, hung on a sandbar in mid lagoon for a day of unremitting roasting. So, M. marks water heights, faithfully, carefully, patiently, and precisely each fifteen minutes, for the two hours from five to seven hours since the previous peak. Life on a tropical atoll can involve more than lolling naked in the sun, even if not much more. No sun today at all, however—overcast sky, muggy air, and a little rain.

### 5 January

Last evening's goat watch provided us with more and closer contacts than all previous watches combined. It was hard to stop for supper. At midnight, the watch was inadvertently resumed when we were roused by odd moaning noises, quit unlike any sound we'd heard. A bright full moon revealed eight goats outside ou tent, apparently conversing in an alien caprine dialect. A few hours later, we aros to start the day, startled at an apparent sunrise in the west-northwest. The true su was cloud-occluded and the full moon, to the north of where the sun normally set fully disoriented us. We still don't understand it.

Our hike east to Cinq Cases (necessitated by the death of both Primus stove and the exhaustion of supplies) began with unsuccessful efforts to rescue a ma rooned green turtle; she was too heavy for us and we had to leave her with hope that clouds and rain would protect her till the next tide. Then came two hours o agonizing balancing on sharp champignons or stumbling through loose coral o deep, soft sand. The occasional stretches of turf were just enough reminder o normal walking to add frustration. The last three hours were smoother, though a ill-designed and badly fitting pack made for maximum discomfort. We did reac Cinq Cases, just an hour after what was to have been our relief party. Engin troubles had delayed them, which was why they never reached us at Dune Jea Louis. It's crowded here, and the camp looks like an impoverished shantytowr Can't wait to get away. The immediate area is a wasteland—coral cliffs and sea t the east, and a vast expanse of largely barren, heavily fragmented, and tumbled loose coral to the west. It's three quarters of an hour to the mangroves and th areas more inhabited by animals, but we will go shortly. The one amenity is th desk from whence this comes—built within the low enveloping branches of Guettarda tree, and shared by dozens of tortoises.

### 6 January

Early up again for a spiritless jog along the coast—so different from Dun Jean Louis. We felt most unambitious, a feeling accentuated by the hot, sticky ai However, we also felt the need for a quiet breakfast, so with a brief bath in the se spray (tide too high and surf too rough for immersion) we returned to our tea an ration biscuits before the others arose. Then we wandered in search of goats. Las night three had appeared close to camp at supper and we spotted others early a.m. on the champignon. But on this walk there was no sign. Lots of brackish pools around which birds teem by the hundreds, as tame as ever. The exception was quartet of flamingos that flew as we approached—a grand view, their long neck outstretched, crimson flashing beneath their wings. Curiously, no one has yet dis covered whether they ever nest here. Perhaps new ones simply periodically appea from distant Africa.

The ample supply of relatively fresh (though foul-smelling) water may ex plain why the incidence of twins (among goats) is so much higher here than a other sites. Lactation would be less of a problem.

The relative openness of the interior vegetation also makes spotting animal easier, so we did finally discover where some animals spend the day—under shad trees, of course, but until today the question of which trees and where had elude us. Today's find was due to an exceptionally favorable wind from the interio which allowed us to track by scent. It is far the best method in this terrain, where to raise one's eyes from the ground, if only for a second, is a sure guarantee of stubbed toes, a wrenched ankle, or worse.

### 8 January

I rose at three a.m., running by moonlight, north to Pt. Hodoul in order to count coastal goats. By five a.m. the turf gave way to coral and I was reduced to a slow stumble. By five-thirty it was a panicky scramble. Only at the very edge of the ledge could a way be forced through the shrubs. Shoes were in ribbons, skin shreds. But just at daybreak I rounded a corner and the dramatic sight of Terre Cedro greeted me. A perfect one-hundred-meter crescent of whitest sand, translucent blue-green water licking the shore, eight ibises patrolling among the cool grove of casuarinas that shaded the sand, and a pair of greens cavorting in the gentle surf (probably like a water bed). I rested only ten minutes, as I had neither water nor food and needed to return before the day's heat set in. The return was a hard slog inland, by sun compass, due south to Basin Flament, then southeast. Some of the way was open, but much was swampy or, worse, thick brush that tore off the remnants of clothes and a few, till then intact, patches of skin. But I did make it, reaching water, food, and a bed by ten.

### 9 January

My jaundiced attitude to the contrary, I had to concede our barbecued billygoat to have been a success. As if in deference to my qualms about eating animals, Harry chunked the meat into anonymous mouth-sized cubes, barbecued them to a crisp on an open wood fire, then mixed them into a peppery onion sauce. Just great, and topped off with rhubarb and strawberries (tinned) over rice. Sybaritic end to a long day.

### 10 January

This wasn't planned to be a writing day, but weather has overruled all. Rain, heavy, unremitting, wet, and cold. We began lazily and late, a slow and short run, then a stumble with pack-sore backs, through the mangrove swamp bordering the lagoon creek where our dinghy *My Fanwy* (pronounced "My Fanny," Welsh name) was anchored. Then, a tedious polling against the tide, finally a ninety minute ride across the lagoon to lovely Middle Camp, our first Aldabra campsite. We stayed out but long enough for tea, then headed west to Anse (beach) Malabar, stopping once, briefly, to shed clothes and gasp at a school (twenty or more) of sharks snorkeling about in the shallows just offshore. The two-and-one-half-hour walk was on the coral ledge above the sea, abominably tiring with the constant ups, downs, dodges, and stumbles the coral imposed, but a trail had been cut through the bush and the seaward view was grand.

Then came those long delayed rains, refreshing at first, then an annoyance, finally, as the wind arose and our temperatures fell, a pain. We reached our camp—on a very lovely crescent beach, one of the very few on the North Shore, thatched wall, tin-roofed, and well-supplied. Except—no can opener and a broken seal on the Primus pump. Grr— We were desperately cold and hungry, so after demolishing several jars of mincemeat, jam, peanut butter, mango chutney, and ration biscuits, which really stimulated the salivary flow, we pried cans open with a knife, sealed the Primus with lard, and cooked tea, sausage, and carrots for a main course. Our bedding got wet, of course, but the gas lantern works well and seems to be drying them. Nothing to do now but sit out the rain—though it could go on for days. Cheerful thought. My shoes are so torn up, however (they've gone at a rate of one pair per week, which statistics ignore the fact that on sand we're barefoot and in water or grass use plastic sandals), that I can't really expect to get further than back to Middle Camp, a trip planned for late tomorrow.

Bedtime now, six hours later, but not without some drama. The heavy rain continued until sundown, so the goat watch was limited to the hut, but enlivened by a newborn kid ten meters distant. Periodic approaches by two adults let us believe it would soon leave, but when it was still bawling and lonely at 2100, we grew softhearted and brought him in for a feed. He's back on his natal ledge now, quiet, but hopefully will still be found by his mother. If not, guess what lies ahead.

Our site is noisy in other ways, too. Surf, of course, plus a whooshing blowhole; then there's a group of boobies roosting nearby (a booby hatchery) and finally a veritable army—dozens upon dozens of claw-clicking, squabbling, investigating—hey, bring our shoes back!—burgher crabs, enormous twenty-cm. diameter, some of them. And, of course the mosquito symphony. Time to duck under the nets!

### 11 January four a.m.

We lost the battle with bugs and crabs—the former biting, the latter swarming over the hut, seizing clothes, dishrags, rattling saucepans. The end came when desperate howls from the kid still alone in the bush led to the discovery it was being eaten alive by the beasts. M. gingerly rescued it and we've just completed a second feed. The kid is in sorry shape, but so are we for that matter. If only it stays dry.

### Nine a.m.

It hasn't. It's ironic, since we'd have been delighted with the water earlier. At least one problem has been unexpectedly solved. Ten minutes ago we heard a dry bleat. We prodded our kid to wakefulness, he replied, and minutes later his mother appeared, walked to within three meters of us, nuzzled him, and the two trotted off together. Incredible. Where had she been since his birth?

### Ten-thirty a.m.

Overcast persists, but only an occasional raindrop wets our bare skin as we lounge on the beach, where the breeze has now picked up enough to clear away gnats and mosquitoes. Another afternoon of rain seems likely, but our walking will not be complicated by an orphaned kid, as we had expected until a couple of hours ago. What a night. The burgher crabs here are the biggest yet seen on Aldabra and more numerous than we've ever seen. There were six tonight in the hut at a time and more every few steps outside. Dune Jean Louis had one that would come into the hut, but rather timidly because Meg threw him out with regularity. Otherwise, we have seen this type mainly on walks scuttling out of the way or just standing with claws raised defensively. They are not aggressive, which I had expected them to be; they didn't grab the stick we used to shove them out of the hut, so we got bold enough to use shoes, but not bare feet. After dark, the noise they make foraging around the hut and squabbling among themselves sounds like a team of clumsy midgets assigned to tear the hut down. Our camp cots were high enough that we didn't really fear their climbing in with us, but many unsettled thoughts accompanied our dozing as we listened to the crabs clank about, and heard the occasional lonely bleat of the kid outside.

### 13 January

One-month anniversary! We began to feel (and resemble) old hands running over the champignon, tan, slim, scratched, bruised and torn. Yesterday's fishing expedition was frustrated first by rock-bound hooks, then a one-and-one-half-meter white-tipped shark that took the line. With the remnants and one other hook P. retired to the lagoon side. There he was met, literally, by nine groupers, lined up side by side like so many trained seals. They rose in unison as the bait was lowered, mouth protruding from the water, almost drooling. But we didn't want groupers. After all of the small ones had twice impaled themselves, been hauled out, freed, and tossed back in, P. gave up.

We were up late with the goats, but failed in our efforts to find a sleeping group that we could capture for ear tagging. Today will be an around-the-clock watch in shifts. P. started the first round and now while I'm having "coffee" (not coffee, but a brew) Meg is preparing to go for her shift as soon as she adjusts the bandages on her other foot. It proved rather a blank day, goat-wise. Total seen: one. Total contact time: one-tenth of a second. We'll try again just before sundown. P. spent the early p.m. constructing a proper windproof fireplace and oven that we tested with a barbecue—sausage, mushroom (tinned), and coconut (fresh)—plus "dampers," flour and water paste rolled over sticks and browned. Quite satisfactory. Tomorrow we'll try our oven, though the bread will have to be unleavened. P. may be able to go fishing, too. No time today, at least not while the tide was out. Fish don't bite at high tide. Bread and fish, with fresh coconut for dessert.

### 14 January

Last night was jackpot night. We had lots of goats to watch from 1700 to dark, and shortly thereafter nabbed a doe and kid, both bedazzled by the bright spotlight Meg wore on her head. While we were measuring and tagging them, the kids bawling all the time, a trio of others appeared and considerately waited their turn. That makes thirteen ear-tagged at Middle Camp, which together with idiosyncratically marked individuals means about one-half to two-thirds of the local population of the Middle Camp site can be named. This morning, as might have been expected, there wasn't a goat to be seen.

*15 January*
Tide's coming in (1100 hours). Now we're all obsessively watching for boats; even Meg spontaneously complained this a.m. of the burden of uncertainty.

Last evening's goat watch produced lots of visual contact, but no catches. Tonight we'll delay our start until 2100, in hopes they'll be thoroughly asleep by then and not see us coming. P. skipped this a.m.'s watch in favor of fishing. We'd almost no protein for the past couple of days. P. very promptly caught a fair-sized (ten- to twelve-kg) coral eater, related to our turbots. It was the shape of an angelfish and multihued, with teeth akin to a dogfish. Little wonder the last one chewed the hook in half. The skin was as tough as leather, but the flesh was easily separated from it and the skeleton to produce filets the color and consistency of beefsteaks. The fires have just been banked to bake M.'s loaf (risen this time, thanks to the sponge she'd set aside last night and which seems to have ripened nicely even without baker's yeast; we'll soon see). The fish will be coal-broiled when the bread is done, to be garnished with a tomato-onion sauce from the Primus. Sounds Lucullan, doesn't it?

We have begun, in shifts, to read Doris Lessing's *The Golden Notebook*. We've not been particular enthusiasts of Lessing's fiction, which is more a reflection upon our parochialism than upon Lessing, but the preface has certainly touched a nerve. M. asked why a writer should explore seemingly trivial emotional or personal issues. This is an analogue of the questions of how we justify much of science. Our work here, for instance, has limited relevance to current social problems. At best, it may help us to preserve the unique biology of Aldabra, but the Ministry of Tourism of the Seychelle's may bring that to naught in any case, if preservation brings hordes of tourists. Is this purely a personal exercise, at the level of other private hobbies? We have never believed so, though to articulate a rationale for this sort of science is another matter. Lessing provides some clues.

Our meal lacked only a chilled Moselle to be awarded at least three stars.

We have one more trip planned here—a one-day excursion at Polymnie, an island we've only seen from boatside. Polymnie is of particular interest because it lacks goats and tortoises, hence an analysis of its soil and vegetation should reveal the impact of the grazers. P. is particularly inclined to the view that the goats are a major factor in the redistribution of nutrients, as they feed in one area (the plateau) and defecate in another (the coastal casuarina or tortoise-grass plots). The tortoises, on the other hand, merely recycle stuff, as they feed and void in one place. Hence, it seems likely that the health of the tortoise population depends on there being some goats. But, how many is that? Why doesn't Dune Jean Louis have more?

No water shortage now, but, for that matter, no reason to use the showers, either. Not even laundry to wash. Our dress (we do wear clothes at Station) consists of plastic sandals and swimsuits. For dinner, pants and shirt, more for mosquito protection than formality. The regulars use *kee-quoys* instead, an East African unisex garb. It consists of a one-and-one-half-by-one-meter rectangle of brightly patterned ultrasoft cotton that is wound around the waist (or shoulder) and then turned under. It's worn Scottish style (nothing under), so it is cool and comfortable, if inhibiting of vigorous movements. *Kee-quoy* accidents among newcomers are a frequent cause of merriment.

### 24 January

Gray day—both externally (rain) and figuratively. Added to the problems posed by the Nordvaer's nonappearance is "native unrest." The laborers whose families are in Mahé have a hard lot; though they're still better off than their unemployed compatriots in Mahé. It is a poor island at best.

Paul, the new director, is just out of school and inexperienced, and thus has been subject to continued test by the workers. He has been thoughtful, firm, and fair, though inevitably has made some mistakes and lost the respect of some. Yesterday, the Seychellois men asked, then cajoled, finally threatened, in efforts to obtain alcohol. The long-standing (though often bent) rule being that it's sold to them only at ration-distribution time on Saturday. Today, most of the men are on strike, and several have announced they will leave with the Nordvaer. Unhappily, the latter group includes several of the oldest, most reliable, and competent workers (also Harry, sad to say). All this could still pass like the rain shower that's now waning. Meantime, no work gets done on Station, research almost ceases (no boat pilots), and tempers fray. One of the several ironies is that had Paul conceded the alcohol Sunday evening, when it was requested, everyone would have almost certainly been too drunk to work today anyway. Tropical doldrums. Oh yes, the Station's chickens all seem to be dying just now. I suspect Newcastle disease. Jim's infected foot (sea urchin spine) is worse. No radio contact has been possible for three days running. Sun, where are you?

### 25 January

Sullen sky, silent radio, morose Seychellois, but by midmorning we were back in our private world at Dune Jean Louis. The remainder of the now brighter morning was spent domestically—erecting tent, reading rain gauge, packing supplies, and hauling fresh sand for the hut floor.

François and P. went pool fishing, François using hands to trap fish under rocky ledges rather than Harry's *pongo*-bashing technique. Finesse. And we got eight. Then a run on the first hard surface we've had since our last trip here. M.'s eight-hundred-meter "track" is around the tortoises.

Now we are on our lookouts, M. on the dune and P. on the tower, hoping for goats. We have an idea as to what has helped control the goat populations. Daytime is when the goats graze in the Pemphis, and then they can't see one another. This would interfere with if not altogether preclude the development of extensive social relations and a stable hierarchy. At night, the goats must retreat to the coastal strip to sleep, and there they come in contact. But the absence of a stable hierarchy, especially in the care of males, leads to repeated assaults on receptive females instead of their domination by the top male. Meg has seen many chases of females by several randy males, chases where violence has led to serious injury or death of the female. This would explain the lopsided sex ratio (it favors males) and a population growth rate well below the maximum. At the open site at Cinq Cases this situation should not obtain. We'll see. It does appear that there are significant demographic differences from one island to the next.

### 26 January

No goats at all. Leaving Dune Jean Louis was difficult for several reasons; goatless, nostalgic, and our boat moored too high up the stream. Yesterday's tide was three meters and today's, we neglected to note, was only two and two-thirds meters, so our boat stayed dry on its side. It took much heaving and straining before we were able to lodge some rollers beneath her, and then came an arduous foot-cutting slog up the damp channel. Even afloat, the engine had to be half-masted and we arrived at Station on the last few cm. of freeboard.

### 30 January

Late last night, amidst driving rain and rising winds, the Nordvaer appeared. Low tide at dawn. The latter didn't produce water enough to move over the reef until late morning, so it was midafternoon before supplies had been offloaded (some en route for eleven months) and our gear stowed. Heavy surf and swells made it a very precarious business, not the least part being the boarding of several infirm Seychellois, including an elderly cardiac case and an immense blimp of obesity, plus several youngsters. A crowded ship.

POSTSCRIPT. The stormy voyage to Mombasa was a horror for all but P., who must have been a water sprite in previous incarnations. Even the captain was seasick.

As for Aldabra, it is now fully part of the Seychelles, though it remains a wildlife refuge, receiving support from the Royal Society and the International Wildlife Conservation groups.

Meg, now Dr. Margaret Gould-Burke, has made several repeat visits to Aldabra and has good evidence to support her view that for the goats, the important demographic factors are not uniform, but site-specific; water shortages at one locale limit reproduction, and the unstable social organization may play a role at another. Control methods, if and when needed, will need to be suited to the particular situations.

**PK watching for goats**

# Works Cited

Allee, W. C. 1938. *Cooperation among animals.* New York: H. Schuman.

Allen, G., et al. 1975. Against sociobiology. *N.Y. Review of Books,* November.

Alvarez-Buylla, A., J. Kirn, and F. Nottebohm. 1990. Birth of projection neuron in adult avian brain may be related to perceptual or motor learning. *Sci.* 249:1444–46.

Antonovics, J. 1987. The evolutionary dys-synthesis: Which bottles for which wine? *American Naturalist* 129:321–31.

Banks, E. 1985. Warder Clyde Allee and the Chicago school of animal behavior. *J. Hist. Behav. Sci.* 21:345–53.

Bateson, C. M. 1972. *Our own metaphor.* New York: Alfred A. Knopf.

Bateson, G. 1960. Some nineteenth-century notions of evolution. Unpub. MS.

———. 1972. *Steps to an ecology of mind.* San Francisco: Chandler.

Bäumer-Schleinkofer, A. 1992a. Biologie unter dem Hakenkreuz. *Universitas* 1:48–61.

———. 1992b. *NS-Biologie und Schule.* Frankfurt: Peter Lang.

Beach, F. A. 1948. *Hormones and behavior.* New York: P. B. Hoeber.

———. 1950. The snark was a boojum. *An. Psychol.* 5:115–24.

———. 1955. The descent of instinct. *Psych. Rev.* 62:401–10.

Bischoff, N. 1991. *Gescheiter als alle die Laffen.* Hamburg: Rasch & Rohring.

Brower, L. 1969. Ecological chemistry. *Sci. Amer.,* February, 22–30.

Budnitz, N. 1978. Feeding behavior of *Lemur catta* in different habitats. In *Perspectives in Ethology III,* edited by P. Bateson and P. Klopfer, 85–108. New York: Plenum Press.

Burkhardt, R. W., Jr. 1981. On the emergence of ethology as a scientific discipline. *Conspectus of History* 1:62–81.

Buss, L. W. 1987. *The evolution of individuality.* Princeton: Princeton University Press.

Caplan, D., A. P. Lecours, and A. Smith. 1984. *Biological perspectives on language.* Cambridge: MIT Press.

Carpenter, R. 1964. *Naturalistic behavior of nonhuman primates.* University Park: Pennsylvania State University Press.

Colinvaux, P. 1980. *The fates of nations: A biological theory of history.* New York: Simon and Schuster.

Collias, N. 1956. Analysis of socialization in sheep and goats. *Ecology* 37:228–39.

Crook, J. H. 1960. Nest form and construction in certain West African weaver birds. *Ibis* 102:1–25

———. 1966. Gelada baboon herd structure and movement. *Symp. Zool. Soc. Pond.* 18:237–58.

Crook, J. H., and J. S. Gartlan. 1966. Evolution of primate societies. *Nature* 210:1200–1203.

Darling, F. 1937. *A herd of red deer.* Cambridge: Cambridge University Press.

Deichmann, U. 1992. *Biologen unter Hitler: Vertreibung, Karrieren, Forschung.* Frankfurt: Campus Verlag.

Dethier, V. 1976. *The hungry fly.* Cambridge: Harvard University Press.

Dewsbury, D.A. 1989. A brief history of the study of animal behavior in N. America. In *Perspectives in Ethology VIII,* edited by P. Bateson and P. Klopfer, 85–122. New York: Plenum.

Dobzhansky, T. 1967. *The biology of ultimate concern.* New York: New American Library.

Durant, J., ed., 1985. *Darwinism and divinity.* Oxford: Basil Blackwell, .

Durkheim, E. 1975. *Durkheim on religion: A selection of readings with bibliography.* London: Routledge and Kegan Paul.

Easter, S. S., et al. 1985. The changing view of neural specificity. *Sci.* 230:507–11.

Edelman, G. M. 1987. *Neural Darwinism.* New York: Basic Books.

Ehle, J. 1965. *The free men.* New York: Harper & Row.

Eialm, D. and I. Golani. 1988. The ontogeny of exploratory behavior in the house rat (Rattus r.): The mobility gradient. *Develop. Psychobiol.* 21:679–710.

Evans, E. P. 1906. *The criminal prosecution and capital punishment of animals.* London: Faber and Faber. Reprint.

Fairchild, L., D. Rubenstein, S. Patti, and P. Klopfer. 1977. A note on seasonal changes in feeding strategies of mixed and single species flock. *Ibis* 119:85–87.

Falk, D. 1990. Brain evolution in *Homo:* The "radiator" theory. *Behavior and Brain Sciences* 13:333–82.

Fausto-Sterling, A. 1998. Animal models for the development of human sexuality. *J. Homosex.* In press.

Feyerabend, P. 1978. *Science in a free society.* London: NLB.

Folley, S. and G. Knagg. 1965. Levels of oxytocin in the jugular vein blood of goats during parturition. *J. Endocrinol.* 33:301–16.

Golani, J. 1973. Non-metric analysis of behavioral interaction sequences in captive jackals. *Behaviour* 44:89–112.

———. 1992. A mobility gradient in the organization of vertebrate movement. *Behav. and Brain Sciences* 15:249–308.

Goldschmidt, R. 1955. *Theoretical genetics.* Berkeley: University of California Press.

Gottlieb, G. 1970. *Development of species identification in birds.* Chicago: University of Chicago Press.

Gould, S. J. 1981. *The mismeasure of man.* New York: W. W. Norton.

Griffin, D. 1958. *Listening in the dark.* New Haven: Yale University Press.

———. 1976. *The question of animal awareness.* New York: Rockefeller University Press.

Gubernick, D., and P. Klopfer. 1981. *Parental care in mammals.* New York: Plenum.

Hagen, J. 1992. *An entangled bank.* New Brunswick, N.J.: Rutgers University Press.

Hailman, J. P. 1967. *The ontogeny of an instinct.* Behav. suppl. Leiden: E. J. Brill.

———. 1982. Evolution and behavior: An iconoclastic view. In *Learning, development, and culture,* edited by H. C. Plotkin. New York: Wiley.

Hardin, G. 1956. The meaninglessness of the word "protoplasm." *Sci. Monthly* 82:112–20

Heinroth, O. 1910. *Beitrage zur Biologie, namentlich Ethologie und Physiologie der Anatiden.* 5 Internat. Ornith. Kong. Verh. 5:589–702.

Hess, E. 1973. *Imprinting.* New York: Van Nostrand Reinhold.

Hinde, R. 1966. *Animal behavior.* New York: McGraw-Hill.

Ho, M.-W. 1988. On not holding nature still: Evolution by process, not by consequence. In *Evolutionary processes and metaphor,* edited by M.-W. Ho and S. Fox. New York: Wiley.

Ho, M.-W., and P. Saunders. 1984. *Beyond Neo-Darwinism.* New York: Academic Press.

Ho, M.-W. , and S. W. Fox. 1988. *Evolutionary processes and metaphors.* New York: Wiley.

Hochbaum, A. 1955. *The travels and traditions of waterfowl.* Minneapolis: University of Minnesota Press.

Honoré, E., and P. Klopfer. 1990. *A concise survey of animal behavior.* San Diego: Academic Press.

Hooper, S., and M. Moulins. 1989. Switching of a neuron from one network to another by sensory-induced changes in membrane properties. *Sci.* 244:1587–89.

Hutchinson, G. E. 1930. Two biological aspects of psycho-analytic theory. *International J. of Psychoanal.* 11:83–86.

Huxley, J. 1942. *Evolution: The modern synthesis.* London: Allen & Unwin.

Jaynes, J. 1969. The historical origins of "ethology" and "comparative psychology." *Anim. Behav.* 17:601–6.

———. 1976. *The origin of consciousness and the breakdown of the bicameral mind.* Boston: Houghton Mifflin.

Jolly, A. 1966. *Lemur behavior.* Chicago: University of Chicago Press.

Kalikow, T. J. 1980. Die ethologische Theorien von Konrad Lorenz: Erklürung und Ideologie, 1938 bis 1943. In *Naturwissenschaft und Technie im Dritten Reich,* edited by H. Mehrtense and S. Richter. Frankfurt: Suhrkamp.

———. 1983. Konrad Lorenz's ethological theory: Explanation and ideology, 1938–1943. *J. Hist. Biol.* 16: 39–72.

Kettlewell, H.B.D. 1961. The phenomena of industrial melanism in the Lepidoptera. *Ann. Rev. Entomol.* 6:245–62.

Kimura, M. 1983. *The neutral theory of molecular evolution.* New York: Cambridge University Press.

Klopfer, P. 1959. An analysis of learning in the Anatidae. *Ecol.* 40:90–102.

Klopfer, P. 1969. Instincts and chromosomes: What is an "innate" act. *American Naturalist* 103:556–60.

———. 1973a. *Behavioral aspects of ecology.* Englewood Cliffs, N.J.: Prentice-Hill.

———. 1973b. Does behavior evolve? *Annals N.Y. Acad. Sci.* 223:113–25.

————. 1977. Social Darwinism lives: Should it? *Yale J. Bio. Med.* 50:77–84.

————. 1982. Mating types and human sexuality. *BioSci.* 32:803–6.

Klopfer, P., and H. Hailman. 1967. *An introduction to animal behavior: Ethology's first century.* Englewood Cliffs, N.J.: Prentice-Hall.

Klopfer, P., and J. Podos. 1998. Behavioral ecology. In *Comparative Psychology: A Handbook,* edited by G. Greenberg and M. Haraway. New York: Garland Press. In press.

Klopfer, P., and J. Polemics. 1988. Have animals rights? *J. Elisha Mitchell Soc.* 104:99–107.

Klopfer, P., and N. Budnitz. 1990. Fixed-action patterns and neural Darwinism. *J. Orn.* 131:97–99.

Klopfer, P,. and R. MacArthur. 1958. North American birds staying on board during an Atlantic crossing. *British Birds* 51:358.

Klopfer, P., M. Klopfer, and J. Etermad. 1981. Girls and horses: A sex difference in attachments. *J. Elisha Mitchell Soc.* 97:1–8.

Koenig, O. 1988. *Wozu aber hat das Vieh diesen Schnabel?* Munich: Piper.

Koestler, A. 1964. *The act of creation.* New York: Macmillan.

Krebs, J. R., and N. B. Davies. 1978. *Behavioral ecology.* Oxford: Basil Blackwell.

Krebs, J. R., and S. Sjölander. 1992. Konrad Zacharias Lorenz. In *Biographic Memoirs of Fellows of the Royal Society,* 211–28. Cambridge: Cambridge University Press.

Kubie, L. S. 1953/1956. Some unsolved problems of the scientific career. Offprint from *Amer. Sci.,* October 1953 and January 1954.

Kuhn, T. S. 1970. *The structure of scientific revolutions.* Chicago: University of Chicago Press.

Lack, D. 1943. *Life of the robin.* London: H. F. and G. Witherby Co.

————. 1954. *The natural regulation of animal numbers.* New York: Oxford University Press.

————. 1956. *Swifts in a tower.* London: Methuen.

Lashley, K. 1929. *Brain mechanisms and intelligence.* Chicago: University of Chicago Press.

Latour, B., and S. Woodgar. 1986. *Laboratory life: The construction of scientific facts.* Princeton: Princeton University Press.

Lederer, S. E. 1992. Political animals: The shaping of biomedical research literature in twentieth-century America. *Isis* 83:61–79.

Lehrman, D. S. 1953. A critique of Konrad Lorenz's theory of instinctive behavior. *Quart. Rev. Biol.* 28:337–63.

Lerner, R. M. 1992. *Final solutions.* University Park: Pennsylvania State University Press.

Levins, R., and R. Lewontin. 1984. *The dialectical biologist.* Cambridge: Harvard University Press.

Leyhausen, P. 1991. Review of *Leben und Werk eines grossen Naturforschers* (1990), by F. M. Wutekis. *Ethol.* 89:344–46.

Lorber, T. 1980. Is your brain really necessary? *Sci.* 210:1232–34.

Lorenz, K. 1940. Durch Domestikation verursachte Störungen arteigenen Verhaltens. *Z. Angew. Psychol.* 59:2–82.

————. 1943. Die angeborenen Formen Möglicher Erfahrung. *Z. Tierpsychol.* 5:235–409.

———. 1953. Über angeborene Instinktformeln beim Menschen. *Deutsche Med. Wochensch.* 78:1566–69, 1600–1604.

———. 1956. The objectivistic theory of instinct. In *L'instinct dans le comportement des animaux et de l'homme,* 51–76. Paris: Maison et Cie.

———. 1974. Analogy as a source of knowledge. In *Les prix Nobel en 1973,* 185–95. Stockholm: The Nobel Foundation.

———. 1981. *The foundations of ethology.* New York: Springer Verlag.

———. 1992. *Die Naturwissenschaft von Menschen.* München: Piper Verlag.

Lumsden, C. J., and E. O. Wilson. 1984. *Genes, mind, and culture.* New York: Random House.

MacArthur, R., and E. Wilson. 1967. *The theory of island biogeography.* Princeton: Princeton University Press.

Masterman, J. C. 1952. *To teach the senators wisdom.* London: Hodder and Stoughton.

Mead, M., and P. Byers. 1968. *The small conference: An innovation of communication.* Paris: Mouton and Co.

Medawar, P. 1957. *The uniqueness of the individual.* London: Methuen.

Miller, D. B., G. Higinbothom, and C. F. Blaich. 1990. Alarm call responsivity of mallard ducklings: Multiple pathways in behavioral development. *Anim. Behav.* 39:1207–12.

Mitman, G. 1992. *The state of nature.* Chicago: University of Chicago Press.

Motokawa, T. 1989. Sushi science and hamburger science. *Persps. in Biol. and Med.* 32:489–504.

Nelson, P. G., C. Yu, and E. A. Neale. 1989. Synaptic connections in vitro: Modulation of number and efficacy by electrical activity during development. *Sci.* 744:585–87.

Oyama, S. 1985. *The ontogeny of information.* New York: Cambridge University Press.

———. 1988. Stasis, development, and heredity. In *Evolutionary processes and metaphors,* edited by M.-W. Ho and S. Fox. New York: Wiley.

Pederson, et al., eds. 1992. Oxytocin in maternal, sexual and social behaviors. *Ann. NY Acad. Sci.* 652:1–492.

Pickering, A., ed. 1992. *Science as practice and culture.* Chicago: University of Chicago Press.

Podos, J. 1994. Early perspectives on the evolution of behavior: C. O. Whitman and O. Heinroth. *Ethobiology, Ecoglogy, and Evolution* 6:467–80.

Polemics, J. 1984. Three paradigms on the relation of science to society. *So. Atlantic Q.* 83:405–15.

Plotkin, H. C. 1988. *The role of behavior in evolution.* Cambridge: MIT Press.

Regan T. 1983. *The case of animal rights.* Berkeley: University of California Press.

Richardson, J., and D. Livingstone. 1962. An attack by a Nile crocodile on a small boat. *Copeia* 1:203–4.

Roeder, K. 1967. *Nerve cells and insect behavior.* Cambridge: Harvard University Press.

Rose, S., L. Kamin and R.C. Lewontin. 1984. *Not in our genes.* Middlesex, U.K.: Pelican Books.

Rubenstein, D., R. Ridgely, R. Barnett, and P. Klopfer. 1977. Migration and species diversity in the tropics. *PNAS* 71:339–40.

Schmidt-Koenig, K. 1975. *Migration and homing in animals.* Berlin: Springer Verlag.

Schurig, V., and S. van Mourik. 1986. Die Begründung der Zoologischen Verhaltensforschung als "Ethologie." *Biol. Rundschau* 24:197–208.

Sebeok, T. A. 1981. *The play of musement.* Bloomington: Indiana University Press.

Singer, P. 1975. *Animal liberation.* New York: Avon Books.

Skinner, B. F. 1971. *Beyond freedom and dignity.* New York: A. Knopf.

Slobodkin, L. B. 1951. The metaphysical structure of creativity. *Amer. Sci.* 39:303–5.

Smillie, D. 1990. Beyond inclusive fitness theory. Unpublished MS.

Snow, C. P. 1961. *Science and government.* Cambridge: Harvard University Press.

Soal, S. G. 1954. *Modern experiments in telepathy.* New Haven: Yale University Press.

Spalding, D. 1954. Instinct: with original observations on young animals. 1873. Reprinted in *Brit. J. Anim. Behav.* 2:1–11.

Stone, C. 1972. *Should trees have standing?* Los Altos, Calif.: William Kaufman.

———. 1987. *Earth and other ethics.* New York: Harper & Row.

Tembrock, G. 1987. *Verhaltensbiologie.* Jena: Fisher Verlag.

Thesleff, S. 1962. A neurophysiological speculation concerning learning. *Perspect. Biol. Med.* 5:293–95.

Thompson, D'Arcy. 1961. *On growth and form.* Abridged ed. Cambridge: Cambridge University Press.

Thorpe, W. H. 1956. *Learning and instinct in animals.* London: Methuen.

———. 1979. *The origins and rise of ethology.* London: Praeger.

Thresher, R. 1977. Territoriality and aggression in the threespot damselfish: An experimental study of causality. Doctoral diss., University of Miami.

Tinbergen, N. 1951. *The study of instinct.* London: Oxford University Press.

Waddington, C. H. 1966. *Principles of development and differentiation.* New York: Macmillan.

Watson, J. 1913. Psychology as the behaviorist views it. *Psychol. Rev.* 20:158–77.

Watson, J. D. 1968. *The double helix.* New York: Atheneum.

Wecker, S. 1963. The role of early experience in habitat selection by the prairie deer mouse. *Ecol. Monogr.* 33:307–25.

Whorf, B. L. 1941. The relation of habitual thought and behavior to language. In *Language in Thinking,* edited by P. Adams. New York: Penguin.

———. 1950. *Language, thought and reality.* Cambridge: MIT Press.

Wickler, W. 1991. Review of *Gescheiterer als die Laffen* (1991), by N. Bischoff. *Ethol.* 89:342–44.

Wieck, M. 1990. *Zeugnis von Untergang Königsbergs: Ein "Geltungsjude" berichtet.* Heidelberg: Verlag Lamber Schneider.

Wilson, D. S. 1980. *The natural selection of populations and communities.* Menlo Park, Calif.: Benjamin/Cummings.

Wilson, E. O. 1975. *Sociobiology.* Cambridge: Belknap Press of Harvard University Press.

Wright, Q. 1942. *A history of war.* Chicago: University of Chicago Press.

Wuketits, F.M. 1990. *Konrad Lorenz. Leben und Werk eines grossen Naturforschers.* Munich: Piper.

Young, H. J., and M. L. Stanton. 1990. Influence of environmental quality on pollen competitive ability in wild radish. *Sci.* 248:1631–33.

Zippelius, H. M. 1992. *Die Vermessene Theorie: Eine kritische Auseinandersetzung mit der Instinkttheorie von Konrad Lorenz und verhaltenskundlicher Forschungspraxis.* Frankfurt: Vieweg-Verlag.

# Index